Bakhtier Vasiev

Nonlinear Dynamical Systems

De Gruyter Series in Applied and Numerical Mathematics

Volume 12

Bakhtier Vasiev

Nonlinear Dynamical Systems

Modeling in Physical, Biological and Social Sciences

DE GRUYTER

Mathematics Subject Classification 2020
Primary: 34A34, 37C75, 37G05; Secondary: 37N25, 93A30, 15A06

Author
Dr. Bakhtier Vasiev
University of Liverpool
Department of Mathematical Sciences
Peach Street
L69 7ZL Liverpool
United Kingdom
bnvasiev@hotmail.com

ISSN 2512-1820
ISBN 978-3-11-134832-2
ISBN 978-3-11-134833-9 (PDF)
ISBN 978-3-11-134839-1 (E-PUB)
DOI https://doi.org/10.1515/9783111348339

Library of Congress Control Number: 2026936620

Bibliographic information published by the Deutsche Nationalbibliothek
The Deutsche Nationalbibliothek lists this publication in the Deutsche Nationalbibliografie; detailed bibliographic data are available on the Internet at https://dnb.dnb.de.

De Gruyter and Walter de Gruyter GmbH are part of De Gruyter Brill.
www.degruyterbrill.com

Questions about General Product Safety Regulation: productsafety@degruyterbrill.com

Printing: CPI books GmbH, Leck

Contents

Preface

The study of dynamical systems offers a powerful framework for understanding how natural and human-made processes evolve in time. From the motion of a falling object to the growth of a population or the emergence of complex biological patterns, many phenomena can be captured by simple mathematical rules whose consequences unfold in rich and sometimes unexpected ways. This textbook grew out of my long-standing interest in such systems and from many years of teaching mathematical modeling to students across the sciences.

The structure of the book reflects this teaching experience. Chapter 1 is based on my introductory course on Dynamic Modeling for first-year students, where the emphasis is on building intuition and learning how mathematical models arise from real-world questions. Chapters 2–6 draw on material from my third-year course on Nonlinear Dynamical Systems, in which students explore more advanced concepts such as bifurcations, oscillations, chaos, and the geometric structures underlying complex behavior. My aim throughout has been to present the subject in a clear and accessible way, while still conveying the depth and elegance of the underlying mathematics. Readers may also find it helpful to watch the video recordings of the lectures and tutorials, available on my YouTube channel: https://www.youtube.com/@bnvasiev.

My own research has been shaped by questions arising in biology, where nonlinear dynamics and pattern formation play a central role. After graduating from the Moscow Institute of Physics and Technology in 1982 and completing my PhD at the Institute of Biophysics of the Academy of Sciences of the USSR in 1990, I have worked at a number of universities and research centers, including the Institute of Theoretical and Experimental Biophysics of the Russian Academy of Sciences, Samarkand State University, Utrecht University, Ludwig Maximilian University in Munich, the University of Dundee, and the University of Liverpool. Much of my work has focused on propagating waves in excitable media and on the chemotactic aggregation of cells, particularly in populations of *Dictyostelium*. In addition, my research includes studies of the mechanisms underlying morphogenetic pattern formation and cell migration in developing tissues, the dynamics of growing populations of microorganisms, theoretical aspects of human population demography, and questions arising in cliodynamics.

I am grateful to several people who have contributed to the preparation of this textbook. I thank my former PhD student, Dr. Manan'Iarivo Louis Rasolonjanahary, for his invaluable help in typing my handwritten lecture notes and shaping a number of the figures. I am also indebted to William Royce and to my daughter, Sofia Vasieva (students of physics and astrophysics, respectively, at the University of Cambridge), for their careful corrections to the text and their skill in preparing and refining the figures. Finally, I express my deepest gratitude to my wife, Olga Vasieva, whose encouragement and inspiration have accompanied me throughout the writing of this book.

https://doi.org/10.1515/9783111348339-203

I hope that readers will find in these pages both a practical introduction to the analysis of dynamical systems and an appreciation for the elegance and coherence of the mathematical ideas that underpin them. Above all, I hope the book encourages curiosity about the models we build, the assumptions we make, and the remarkable dynamical regimes that arise in simple settings.

1 Modeling dynamical systems

Let us start with the statement that everything we know is changing over time. What seemingly is not changing is also changing, but at a slower pace. Contemporary science is extremely successful in describing and explaining observed changes in many natural processes. This success is achieved due to the widely used scientific method (developed in the 17th century), whereby the phenomenon under consideration is isolated (mentally) from all other processes prior to its analysis. Such an isolated object is called a system, while its mental recreation is a model. Systems evolving over time are called dynamical, as opposed to those which are considered to remain unchanged, or static.

Studies of both statical and dynamical systems involve experimentation as well as the use of mathematics. The state of a dynamical system can be described by a quantity (or a set of quantities) that changes over time. Mathematical modeling provides a way to predict such changes, which can then be compared with observations or experimental data. The modeling is based on a simplification of the real-world problem, so that the latter is converted into a mathematical problem. This can then be solved using analytical and numerical techniques, and the obtained solution can be used to explain and predict the behavior of the modelled natural system.

1.1 Modeling examples

To illustrate how modeling works, we will consider three problems which eventually lead to nondynamical modeling.

Example 1.1. A small rectangular plot of farmland with an area of 216 m^2 is to be enclosed by a fence and divided into two equal plots by another fence, parallel to one of the sides. What dimensions for the outer plot will require the smallest total length of fencing (i. e., the least amount of material to be used for fencing)?

Solution:
Now we need to reformulate the problem so that it is represented by an equation for the dimensions of the plot, and we answer the problem by solving this equation. Let us denote the unknown dimensions by x and y. A graphical representation of the plot, divided by a fence parallel to the x-side, is given in Figure 1.1. The required equation can

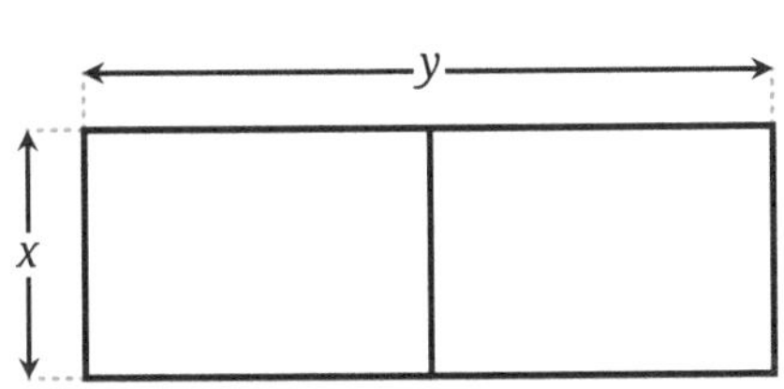

Figure 1.1: Illustration to Example 1.1. The farmland is represented by a rectangle with dimensions x and y. It is divided by a fence parallel to the x-side.

https://doi.org/10.1515/9783111348339-001

be derived by calculating the area,

$$A = xy,$$

and the total length of fencing,

$$L = 3x + 2y,$$

for the given geometrical configuration.

As the area A is fixed, the two unknowns, x and y, are dependent. For any given value of x, we have

$$y = \frac{A}{x},$$

so that

$$L(x) = 3x + \frac{2A}{x}.$$

The requirement for the minimal length of fencing is now reduced to the requirement that the function $L(x)$ has a minimum value, which occurs when the derivative of $L(x)$ is zero as follows:

$$\frac{dL}{dx} = 3 - \frac{2A}{x^2} = 0.$$

This leads to the equation

$$x^2 = \frac{2A}{3} = \frac{2 * 216\,\text{m}^2}{3},$$

which gives $x = 12$ meters, $y = A/x = 18$ meters, and, correspondingly, the total length of fencing is $L = 3x + 2y = 72$ meters.

Note that, to solve this problem, we had to identify a specific function $L(x)$ and observe that the solution reduces to finding the extremum of this function, specifically, its minimum value. In general, mathematical modeling involves four steps:

- Problem description;
- Model formulation;
- Finding the model solution;
- Interpretation of the solution.

The key point is the formulation of a mathematical model. This is a challenging task and can only be learned through practice. A number of examples presented in this textbook are intended to illustrate both the formulation and the analysis of mathematical models.

Example 1.2. A ladder has to be carried around a corridor corner. Assuming the corridor dimensions are a and b, what is the maximum length of the ladder that can pass around the corner?

Solution:
The corner and a ladder of length L, which is carried around the corridor and touches the walls at points P and Q, are shown in Figure 1.2 (left panel). The distance between the corner edge, N, and the ladder is denoted by d. The longer the ladder (i. e., the greater the value of L), the shorter the minimum distance between the ladder and point N becomes. If the ladder is too long, it cannot be carried around the corner because it will simultaneously touch both outer walls and the inner corner (as shown in Figure 1.2 (right panel)).

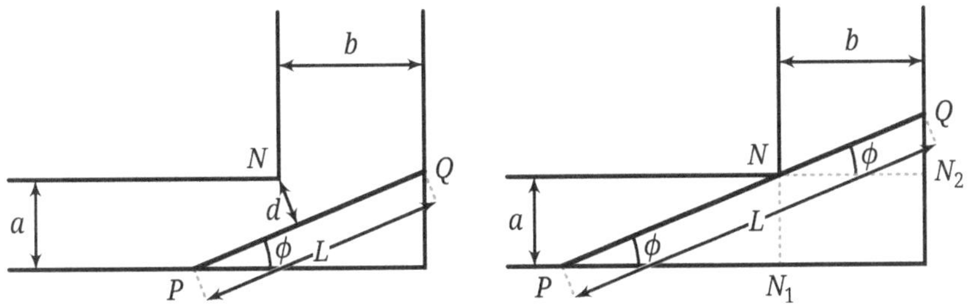

Figure 1.2: Illustration to Example 1.2. Schematic representation of the ladder (of length L) that can be carried (left panel) and cannot be carried (right panel) around the corner. When the ladder is stuck, it simultaneously touches both walls (at points P and Q) and the inner corner (point N). In this position, it forms an angle ϕ with the wall along the corridor of width a.

One way to solve the problem is to find the length of the ladder, L, such that the minimum distance d between point N and the ladder is zero, $d = 0$. Then, any longer ladder would become stuck, while any shorter ladder could be carried around the corner. The length of the ladder that touches point N can be expressed as

$$L = PN + NQ = \frac{a}{\sin\phi} + \frac{b}{\cos\phi}, \tag{1.1}$$

which gives the relationship between the ladder length, L, and the angle, ϕ, as illustrated by the plot in Figure 1.3. As we can see, when the ladder gets stuck, it may be positioned at one of two different angles to the same wall. These angles correspond to the two possible directions in which the ladder can be carried around the corner.

For the shortest possible ladder that touches point N (and just passes around the corner), we require

$$\frac{dL}{d\phi} = 0 \quad \text{or} \quad \frac{d}{d\phi}\left(\frac{a}{\sin\phi} + \frac{b}{\cos\phi}\right) = 0,$$

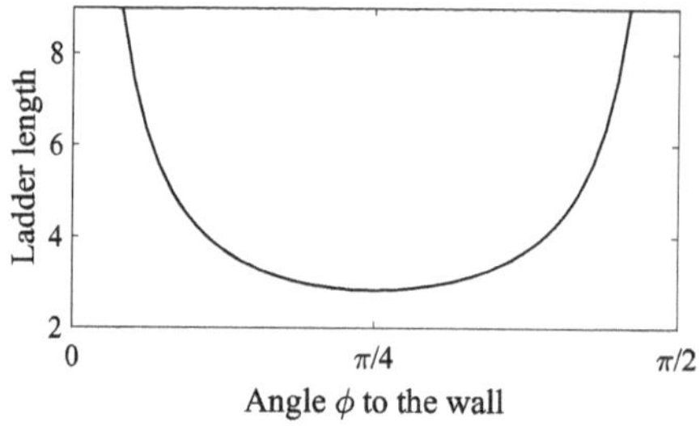

Figure 1.3: The length of the ladder as a function of the angle ϕ at which it becomes stuck in the corner, as given by formula (1.1). Plotted for $a = b = 1$.

which, for the critical angle, gives

$$\frac{a\cos\phi}{\sin^2\phi} - \frac{b\sin\phi}{\cos^2\phi} = 0 \quad \text{or} \quad \tan\phi = \left(\frac{a}{b}\right)^{1/3}.$$

To find $\sin\phi$, we use the identity

$$\tan^2\phi = \frac{\sin^2\phi}{\cos^2\phi} = \frac{\sin^2\phi}{1-\sin^2\phi} = \left(\frac{a}{b}\right)^{2/3}.$$

From here, we find both trigonometric functions that appear in Eq. (1.1)

$$\sin\phi = \sqrt{\frac{a^{2/3}}{a^{2/3}+b^{2/3}}} \quad \text{and} \quad \cos\phi = \sqrt{\frac{b^{2/3}}{a^{2/3}+b^{2/3}}}.$$

Finally,

$$L_0 = \frac{a}{\sin\phi} + \frac{b}{\cos\phi} = \cdots = (a^{2/3}+b^{2/3})^{3/2}.$$

If the length of the ladder is greater than L_0, it will get stuck when carried around the corner (the angle ϕ at which it gets stuck is given implicitly by formula (1.1)). If the ladder is shorter than L_0, it can be carried around the corner.

Note that, in order to solve this problem, we derived the function (1.1), and analysis of this function led us to the solution.

Example 1.3. An oil company increases the price of the petrol it sells to a petrol station owner. The petrol station owner now faces a problem. Some of the price increase must be passed on to customers in order to maintain profits. However, if too much is passed on, it may lead to a loss of customers, which would reduce profits. The owner would like to determine exactly what price increase will maximise profits.

Solution:
In order to solve this problem, we need to introduce a few variables and make some assumptions about them. Let us denote

- c: the cost of petrol (per litre) set by the oil company;
- s: the sale price set by the petrol station owner;
- V: the volume of petrol sold (per week);
- p: the profit (per week) made by the petrol station owner.

Then, the weekly profit can be calculated as

$$p = (\text{selling price} - \text{cost price}) \times \text{volume of sales} = (s - c)V.$$

Suppose the oil company increases its price (the cost of petrol) by Δc, and the petrol station increases its sale price by x. The volume of sales will decrease depending on the change in price, so we write $V = V(x)$. We assume that the drop in sales is proportional to the increase in price, so

$$V(x) = V_0 - ax, \tag{1.2}$$

where a is an unknown coefficient of proportionality and V_0 is the volume of sales when the sale price remains unchanged. The weekly profit now depends on both the increase in the oil price by the oil company, Δc, and the increase in the sale price by the petrol station owner, x. It is given by

$$p(x) = [(s + x) - (c + \Delta c)](V_0 - ax).$$

Here we consider the profit $p = p(x)$ as a function of the variable x. Similarly to the previous example, we need to find the value of x that maximizes the profit $p(x)$. We observe that $p(x)$ is a quadratic function, and there are two roots of the equation $p(x) = 0$,

$$x = \frac{V_0}{a} \quad \text{and} \quad x = c + \Delta c - s.$$

Furthermore, $p(x)$ is a concave function (a quadratic with a negative leading coefficient), so its extreme point, located exactly midway between the roots, corresponds to a maximum (see Figure 1.4). Thus,

$$x_{\max} = \frac{V_0 - a(s - c - \Delta c)}{2a}.$$

There is an unknown parameter, a, on the right-hand side of this equation. If the original sale price, x_0, was already optimized, then $x_{\max} = 0$ when $\Delta c = 0$. This condition gives

$$a = \frac{V_0}{s - c} \quad \text{and, correspondingly,} \quad x_{\max} = \frac{\Delta c}{2},$$

meaning that half of the cost increase should be passed on to customers. Otherwise, to determine the parameter a, the station owner must carry out one trial by measuring the

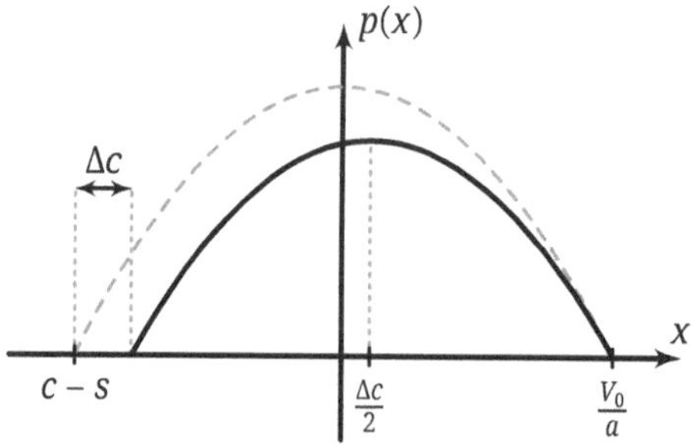

Figure 1.4: The profit $p(x)$ is a quadratic function of the change in sale price x, with a maximum at $x = \Delta c/2$, where Δc is the change in cost. The dashed curve represents the case $\Delta c = 0$.

profit p_1 corresponding to an arbitrary sale price increase $x = x_1$. Then, a can be found from the equation

$$p_1 = (s + x_1 - c - \Delta c)(V_0 - a x_1),$$

which leads to

$$a = \frac{V_0(s + x_1 - c - \Delta c) - p_1}{x_1(s + x_1 - c - \Delta c)}.$$

As we have seen from the considered examples, mathematical modeling can be of great use for various real-life problems. More mathematical modeling examples can be found in a number of textbooks including [1, 2].

1.2 Linear approximation

In order to deal with the problem in Example 1.3, we had to make an important assumption, expressed by Eq. (1.2), that the volume of sale is a linear function of the change in sale price. This assumption is not as arbitrary as it may appear. It is the so-called linear approximation of the function $V(x)$. This concept is so important that we will look at it more closely. But before getting to it, I should introduce the so-called little-o notation.

Definition 1.1. Consider function $f(x)$ defined in the neighborhood of $x = 0$, that is, there exist δ, such that $f(x)$ is defined for any x satisfying $|x| < \delta$. Assume that $f(x) \to 0$ when $x \to 0$. Then, if

$$\lim_{x \to 0} \frac{f(x)}{x} = 0,$$

we then say $f(x)$ is a **little-*o*** of x, or $f(x) = o(x)$.

In other words, $o(x)$ is a class of functions that tend to zero faster than x. $f(x) = x^2$ is an example of such a function

$$\lim_{x \to 0} \frac{f(x)}{x} = \lim_{x \to 0} \frac{x^2}{x} = 0,$$

and therefore $x^2 = o(x)$. Using equality notation in the definition of little-o is common practice, although formally it would be more correct to use the 'belongs to a class of functions sign', as in $x^2 \in o(x)$.

The definition of little-o can be extended to other arguments. For example, we state that $f(x) = o(x^2)$ if

$$\lim_{x \to 0} \frac{f(x)}{x^2} = 0,$$

that is, when $x \to 0$, then obviously $x^2 \to 0$ faster, but $f(x) \to 0$ even faster than x^2.

More information on little-o notation can be found in the textbook [6]. For completeness, we also introduce here a class of big-O functions.

Definition 1.2. If, for the function $f(x)$ defined in the neighborhood of $x = 0$, we have

$$\lim_{x \to 0} \frac{f(x)}{x} = C,$$

where C is a finite number, that is, $0 < |C| < \infty$, then we say $f(x)$ is a **big-O** of x, or $f(x) = O(x)$.

In other words, class of $O(x)$ functions contains those tending to zero in the same pace as x. $f(x) = \sin x$ is an example of such a function

$$\lim_{x \to 0} \frac{f(x)}{x} = \lim_{x \to 0} \frac{\sin x}{x} = 1,$$

and therefore $\sin x = O(x)$.

Now, we are ready to talk about linear approximation of functions.

Definition 1.3. Consider the function $f(x)$ which is defined and continuous in some neighborhood of point $x = a$ (that is, there exist δ such that $f(x)$ is defined and continuous for all x satisfying $|x - a| < \delta$). If $f(x)$ has a derivative $f'(x)$ at point $x = a$, then the function

$$u(x) = f(a) + f'(a)(x - a) \tag{1.3}$$

is called a **linear approximation** of $f(x)$ at point $x = a$.

This means that the functions $f(x)$ and $u(x)$ become almost indistinguishable as $x \to a$. This statement can be expressed as $y(x) \approx u(x)$, when $|x - a|$ is a small number, or by the formula

$$f(x) = f(a) + f'(a)(x - a) + o(x - a). \tag{1.4}$$

To prove this formula, we consider the function $h(x) = f(x) - u(x)$ representing the difference between $f(x)$ and its linear approximation, $u(x)$ given by (1.3). Using L'Hôpital's rule, we get

$$\lim_{x\to a}\frac{h(x)}{x-a} = \lim_{x\to a}\frac{f(x)-u(x)}{x-a} = \lim_{x\to a}\frac{f'(x)-u'(x)}{1} = f'(a) - u'(a) = 0,$$

and thus, $h(x) = o(x)$, which is stated by Eq. (1.4). Note, that the function $u(x)$ represents a tangential line for function $f(x)$ at point $x = a$.

Example 1.4. Find the linear approximation of the function $y(x) = \sqrt{1+x}$ at $a = 0$.

Solution:
According to the formula (1.4),

$$y(x) \approxeq y(a) + y'(a)x = \sqrt{1+0} + \frac{1}{2\sqrt{1+0}}x = 1 + \frac{x}{2}.$$

Plots of the function $y(x) = \sqrt{1+x}$ and its linear approximation are shown in Figure 1.5. In this example as in many other practical cases, the approximation is made at point $a = 0$. Formulas for the linear approximation of the most commonly used functions at $a = 0$ include

$$\begin{aligned} e^x &\approx 1 + x \\ \sin x &\approx x \\ (1+x)^\alpha &\approx 1 + \alpha x \\ \ln(1+x) &\approx x. \end{aligned}$$

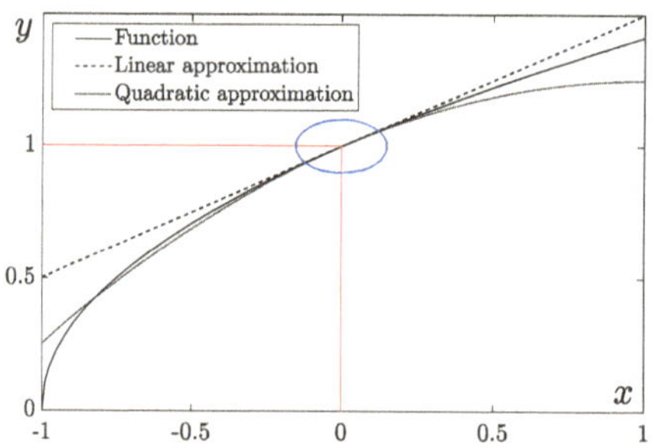

Figure 1.5: Illustration of the linear and quadratic approximations of the function $y = \sqrt{1+x}$ at $x = 0$. In the vicinity of the point $(x, y) = (0, 1)$ (indicated by a blue oval), the approximations are almost indistinguishable from the actual function.

The definition of a linear approximation can be extended to introduce a second-order approximation, which provides an even closer fit to the function being approximated.

Definition 1.4. If the function $f(x)$ is defined and has a continuous first derivative in the vicinity of the point $x = a$, and also has a second derivative at this point, then it can be approximated near $x = a$ by a quadratic parabola, representing **quadratic approximation**, so that

$$f(x) = f(a) + f'(a)(x - a) + \frac{1}{2}f''(a)(x - a)^2 + o((x - a)^2).$$

This equation is proven in the same way as Eq. (1.4), but using L'Hôpital's rule twice. Both the linear and quadratic approximations are particular cases of the so-called Taylor expansion, which can be found in standard textbooks on calculus (for example, [18]).

Example 1.5. Find the value of $\sqrt[3]{130}$ to the second decimal place.

Solution:
The idea here is to use a linear or quadratic approximation for the binomial function $(1 + x)^{\alpha}$. Now,

$$\sqrt[3]{130} = \sqrt[3]{125 + 5} = \sqrt[3]{125\left(1 + \frac{5}{125}\right)} = 5\sqrt[3]{1 + \frac{1}{25}} = 5(1 + x)^{\alpha},$$

where $x = 1/25$ and $\alpha = 1/3$. The first-order approximation, to the second decimal place, is

$$\sqrt[3]{130} = 5\left(1 + \frac{1}{3} \times \frac{1}{25}\right) = 5 + \frac{1}{15} \approx 5.07.$$

To check its accuracy, we can find the second-order approximation

$$\sqrt[3]{130} = 5\left(1 + \frac{1}{3} \times \frac{1}{25} + \frac{1}{3}\frac{(-2)}{3}\frac{1}{2}\frac{1}{25^2}\right) = 5 + \frac{1}{15} - \frac{1}{9 * 125} \approx 5.07.$$

As we can see, the linear approximation is sufficiently accurate in this particular example.

1.3 Dynamic modeling

Mathematical models used for solving various practical problems can fall into various classes. Design models are commonly used in engineering for the construction of buildings, bridges, cars, and aeroplanes. Descriptive models are typically employed in science to analyze natural phenomena such as the motion of planets, the properties of solids, liquids, and gases, and the structure of matter. Of particular interest to us here is so-called *dynamic modeling*: The word "dynamic" refers to evolution over time. It is used in problems involving changes in certain quantities over time, and our modeling will aim to describe or predict those changes.

Definition 1.5. A **dynamical system** is one that is described by variables which change over time, i. e., $x(t), y(t)$ and so on.

The number of variables describing the system is often referred to as the number of its dimensions. This terminology comes from associating the variables describing the system with the coordinates of a point. For example, a one-dimensional system involves finding the position of a particle moving along a line; a two-dimensional system describes a particle moving in a plane, and so on. The variables can be either continuous (with their values represented by real numbers) or discrete (i. e., given by integers). This is especially true of the variable representing time.

Definition 1.6. **Discrete-time dynamical systems**, or **maps**, are commonly described by **difference equations**, which, in the simplest case, take the form

$$u_{n+1} = f(u_n). \tag{1.5}$$

The function $f(u)$ is called the **map generating function** and u_n describes the state of the system after n number of time intervals (e. g., years) starting with some known state (u_0). The solution is represented by a sequence of numbers, $u_0, u_1, u_2, \dots, u_{n-1}, u_n, \dots$, and is called an **orbit**. It is uniquely defined by the function $f(u)$ and the initial value u_0.

Example 1.6. Derive a difference equation modeling the population of a town under each of the following assumptions:

(a) The population increases by 1,000 every year;

(b) The population increases by 1 % every year;

(c) The population increases by 2 % every year and there is a constant annual emigration of 100 individuals.

Solution:

Let us denote by u_n the size of the population in year n. Based on the described changes in u_n, we should define the function $f(u_n)$ such that the model is represented by Eq. (1.5).

(a) Here, we have an increase in u_n by 1,000 every year, i. e.,

$$u_{n+1} = u_n + 1{,}000.$$

The solution is represented by a sequence called an arithmetic progression. Solutions for two different values of u_0 are shown in Figure 1.6A.

(b) In this case, we have an extra 1 %, or 0.01, of u_n added each year:

$$u_{n+1} = 1.01u_n.$$

Now, the solution is represented by a geometric progression. Two solutions corresponding to two different initial sizes of town's population are illustrated in Figure 1.6B.

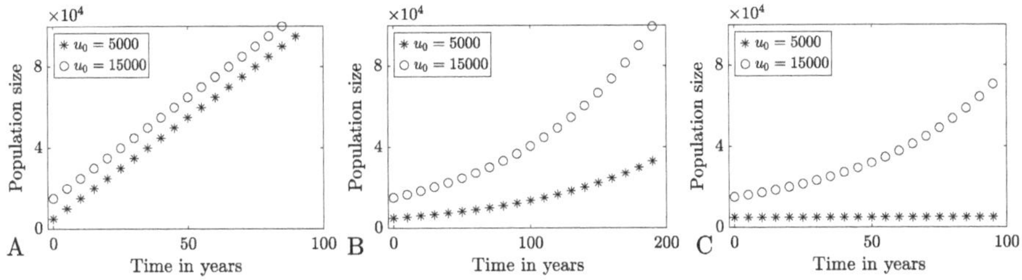

Figure 1.6: Illustration of solutions to Example 1.6. Panel A: solutions to Example 1.6A, represented by arithmetic progressions. **Panel B:** solutions to Example 1.6B, represented by geometric progressions. **Panel C:** solutions to Example 1.6C, showing the stationary solution for $u_0 = 5{,}000$ and the geometric progression represented by the difference $u_n - 5{,}000$ for the solution with $u_0 = 15{,}000$.

(c) Now, we have an extra 2 %, or 0.02, of u_n, and, in addition, a reduction of 100 every year:

$$u_{n+1} = 1.02u_n - 100.$$

The solution is a sort of geometric progression, as illustrated in Figure 1.6C. If the initial town population is 5,000, then the size of the population does not change over time: Each year there is an increase of 2 %, which is 100 people. This, together with the reduction of the same size, results in no net change in the population size. If the initial population is larger than 5,000, then the difference $u_n - 5{,}000$ forms a geometric progression. These statements are illustrated by two solutions for $u_0 = 5{,}000$ and $u_0 = 15{,}000$, shown in Figure 1.6C.

Definition 1.7. Continuous dynamical systems or **flows** (where time is represented by a continuous variable t) are described by **ordinary differential equations** (ODEs).

The modeling of such systems is associated with the identification of the variable (or set of variables), say x, describing its state, and the analysis of the rate of change of this variable. The associated differential equation, typically having the form

$$\frac{dx}{dt} = f(x,t), \tag{1.6}$$

has a solution, $x(t, x_0)$, that depends on the initial value of x, $x_0 = x(t = 0)$, and defines the state of system at time t. The derivative with respect to time is commonly denoted by a dot, $\dot{x}$, which is known as Newton's notation for the derivative.

Example 1.7. At a railway station, customers arrive at the booking office at the rate of 2 per minute and are served at the constant rate R per minute. When the ticket office opens, there are 20 customers waiting. How does the size of the queue at the booking office evolve over time?

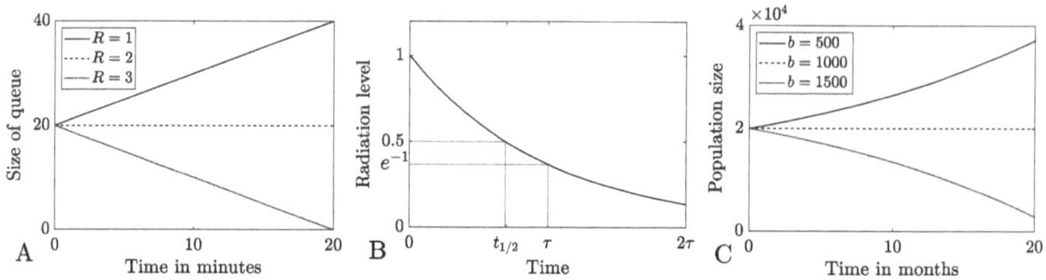

Figure 1.7: Illustration of solutions to Examples 1.7–1.9. Panel A: illustration of the solutions to Example 1.7; depending on the rate of serving R, the queue is increasing, stationary, or decreasing. **Panel B:** solution to Example 1.8, showing the exponential decay of radiation (or the amount of radioactive material). **Panel C:** illustration of the solutions to Example 1.9 for an initial population of $x_0 = 2 \times 10^4$ rats; depending on the rate of extermination b, the population is increasing, stationary, or decreasing.

Solution:
Let $x(t)$ to be a number of customers at time t at the ticket office. The rate of change is given by two arriving customers and R served and leaving (per minute). Thus, the model equation is

$$\dot{x} = 2 - R.$$

Integration yields

$$x(t) = x_0 + (2 - R)t,$$

where $x_0 = 20$ is the initial size of the queue. Solutions for $R = 1, 2$, and 3 are illustrated in Figure 1.7A. When R is small, the queue is growing over time, while, when R is large, it is decreasing. Although the solution is given by the variable $x(t)$, which takes real values, we understand that the number of customers is given by a natural number represented by the integer part of the solution $x(t)$.

The rate of change in the example is constant, and, correspondingly, the size of the queue changes linearly over time. This dynamics is similar to that in the difference equation given by the arithmetic progression (see Figure 1.6A). Also, the solution represented by a geometric progression in a difference equation is closely associated with the solution given by an exponential function in a differential equation.

Example 1.8. It is known that radioactive materials decay due to radiation, and the rate of decay is proportional to the amount of material, x, with certain coefficient of proportionality, a, so that the differential equation describing dynamics of the process is

$$\dot{x} = -ax,$$

where $a > 0$. The constant a can be rewritten as $1/\tau$, so that the parameter τ (called the *relaxation time*) defines the time scale of the process. Thus, the previous equation can be rewritten as

$$\dot{x} = -\frac{x}{\tau},$$

and its integration gives

$$x(t) = x_0 e^{-t/\tau},$$

where $x_0 = x(t = 0)$ is the initial amount of radioactive material. This is the formula for the exponential decay of radioactive material (or the level of radioactivity), as illustrated in Figure 1.7B. At time τ, the amount of material has dropped by a factor of e. The relaxation time τ is related to the well-known half-life, $t_{1/2}$, of radioactive material, which is the time required for a two-fold reduction in radiation (or in the mass of the radioactive material). The equation is

$$x(t_{1/2}) = 0.5x_0 = x_0 e^{-t_{1/2}/\tau} \quad \text{or} \quad t_{1/2} = \tau \ln 2.$$

In Example 1.7, we had dynamics with a constant rate of change, whereas in Example 1.8 this rate depends on the value of the variable, x, describing the state of the system. This is called a functional rate of change, and we will present another example where the rate of change is given by a linear function, $f(x)$.

Example 1.9. A rat population has a functional growth rate of 5 % per month. It is decided to set up an extermination program capable of killing b rats per month. Model the results.

Solution:
Let $x(t)$ be the size of the rat population at time t, with x_0 being the initial size, i. e., $x_0 = x(t = 0)$. Then, the differential equation describing the dynamics of the rat population is

$$\dot{x} = 0.05x - b,$$

where the first term on the right-hand side represents the 5 % increase in the population size and the second term accounts for the effect of the extermination program. This differential equation is solved using the standard method of separation of variables

$$\frac{dx}{x - 20b} = 0.05dt.$$

The integration gives

$$\ln(x - 20b) = 0.05t + c \quad \text{or} \quad x - 20b = Ae^{0.05t},$$

where c is a constant of integration and $A = e^c$. Setting $t = 0$, we get $A = x_0 - 20b$ and, therefore, the solution is

$$x(t) = (x_0 - 20b)e^{0.05t} + 20b.$$

The difference $x - 20b$ increases exponentially if it is positive, or decreases if negative. It remains unchanged over time, i. e., $x(t)$ is constant if $x - 20b = 0$. These three cases are illustrated in Figure 1.7C.

The last four examples have illustrated the use of difference and differential equations in modeling dynamical systems. The same dynamical problem can be solved using both approaches, as will be shown in the next example.

Example 1.10. The temperature of tea in a mug dropped from 100 °C to 90 °C in 1 minute. What will the temperature of the tea be after 5 minutes, assuming the room temperature is 20 °C?

Solution:
Let $T(t)$ be the temperature of the tea at time t. Assume that the rate of change of this temperature depends only on the temperature itself (i. e., a functional rate of change)

$$\dot{T} = f(T).$$

Let us consider the linear approximation of the function $f(T)$ at the point $T_s = 20$ °C—the room temperature; i. e. $f(T) = f(T_s) + f'(T_s)(T - T_s)$. Thus,

$$\dot{T} = f(T_s) + f'(T_s)(T - T_s).$$

Obviously, if the temperature of the tea is equal to the room temperature, it will not change, and therefore $f(T_s) = 0$. Furthermore, the temperature of the tea drops if the room temperature is below that of the tea, which means $f'(T_s)$ is negative. Let us introduce the positive parameter k such that $k = -f'(T_s)$. Thus, we get

$$\dot{T} = -k(T - T_s). \tag{1.7}$$

This equation represents the well-known Newton's law of cooling. Experimental studies have verified that this equation is quite accurate across a wide range of temperatures, in various experimental settings and for many materials. Integration of this equation (see Example 1.9) gives

$$\ln(T - T_s) = -kt + \ln(T_0 - T_s), \tag{1.8}$$

where $T_0 = 100$ °C is the initial temperature of the tea. From here,

$$k = \frac{\ln(T_0 - T_s) - \ln(T - T_s)}{t} = \frac{1}{t}\ln\frac{T_0 - T_s}{T - T_s},$$

and taking into account that the temperature of the tea drops to 90 °C in 1 minute, we find the value of the parameter k as follows:

$$k = \ln \frac{8}{7} \approx 0.1335.$$

Also, from (1.8) we have

$$T = (T_0 - T_s)e^{-kt} + T_s,$$

from which we get

$$T(t = 5) = 80 \times 0.875^5 + 20 = 61\,°\text{C}.$$

To model this problem using a difference equation, we introduce the variable T_n—the temperature of the tea after n time intervals, that is, at time $t = n\Delta t$, where Δt is a certain time interval, say 1 minute. As follows from Eq. (1.7), we should consider the functional rate in terms of the difference $T_n - T_s$:

$$T_n - T_s = (1 - k\Delta t)(T_{n-1} - T_s),$$

from which we obtain:

$$k = \frac{T_0 - T_1}{(T_0 - T_s)\Delta t}$$

to find k, and

$$T_n = (1 - k\Delta t)^n (T_0 - T_s) + T_s$$

to find the temperature after n time intervals. To find k, we use the condition that the temperature drops to 90 °C in one minute, set Δt to one minute, and obtain $k = \frac{1}{8}$. Now, we can find the temperature of the tea after 5 minutes as follows:

$$T_5 = (1 - k\Delta t)^5 (T_0 - T_s) + T_s = (1 - 1/8)^5 * 80 + 20 = 61$$

which coincides with the answer we obtained using the differential equation (1.7).

Modeling dynamical processes does not necessarily require the introduction of dynamic variables; in such cases, we do not consider the modelled object as a dynamical system. We will illustrate this with the following example.

Example 1.11. Traffic lights can be considered a form of obstruction. Suppose we have two roads intersecting at right angles. Cars arrive at a rate of 4 cars per minute at one set of lights, which is green for a time T_1. On the other road, cars arrive at a rate of 3 cars per minute, and the green light lasts for T_2 minutes. Additionally, both lights are red for one minute before the second set of lights turns green, to allow pedestrians to cross. Traffic passes through a green light at a rate of 10 cars per minute. What conditions on T_1 and T_2 should be satisfied to avoid congestion?

Solution:
One way to address this problem is by introducing dynamic variables, say $x(t)$ and $y(t)$, representing the number of cars at each set of traffic lights at time t. However, many problems can be solved more quickly and efficiently by considering the integral characteristics (so-called 'first integrals'). We note that the process is periodic, with period $P = T_1 + T_2 + 1$. Within one period, the number of cars arriving at the first set of lights is $4P$, and the maximum number of cars leaving is $10T_1$. The condition to avoid congestion at this set of lights is

$$10T_1 \geq 4P = 4(T_1 + T_2 + 1) \quad \text{or} \quad T_2 \leq 1.5T_1 - 1.$$

On the plane defined by T_1 and T_2, this condition is satisfied on and below the line $T_2 = 1.5T_1 - 1$, which is shown by the solid line in Figure 1.8.

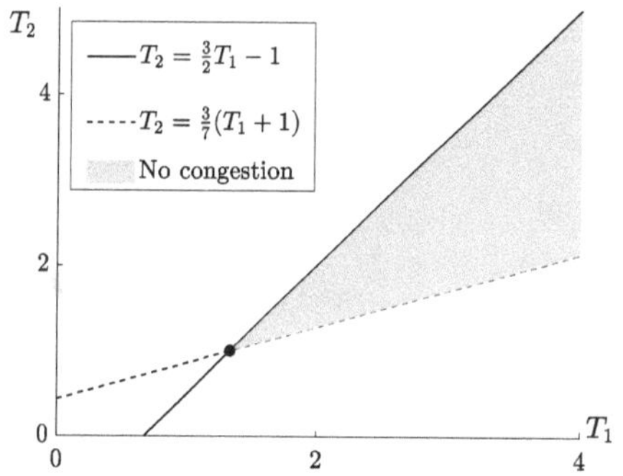

Figure 1.8: Illustration of solution to Example 1.11. The domain on the (T_1, T_2) plane where there is no congestion at the crossroads is shaded in grey. The marked point at the intersection of the solid and dashed lines corresponds to the lowest pair $(T_1, T_2) = (4/3, 1)$ that allows for no congestion.

Similarly, the condition to avoid congestion at the second set of lights is

$$10T_2 \geq 3P = 3(T_1 + T_2 + 1) \quad \text{or} \quad T_2 \geq 3/7(T_1 + 1).$$

On the plane T_1, T_2, this condition is satisfied on and above the line $T_2 = 3/7(T_1+1)$, which is shown by the dashed line in Figure 1.8. Both conditions must be satisfied to avoid congestion at both sets of lights on the crossroad. The domain where both conditions hold is shaded in green in Figure 1.8. The minimal values of T_1 and T_2 in this domain are found from the condition

$$1.5T_1 - 1 = 3/7(T_1 + 1),$$

which gives $T_1 = 4/3$ minutes and $T_2 = 1$ minute.

1.4 Summary

In this chapter we have seen how different modeling techniques can be used to analyze a variety of real-life problems. In particular, we have learned (or recalled) how to formulate and work with both discrete-time and continuous models of dynamical systems. These two modeling frameworks will form the backbone of the material developed in the following chapters.

In Chapter 2, we will focus on one-variable discrete-time models, expressed in the form of difference equations. We will explore the full range of dynamical behaviors that such models can exhibit, from simple convergence to fixed points to more complex oscillatory patterns. Remarkably, we will also see that chaotic dynamics can arise even in these very simple one-dimensional systems.

In Chapter 3, we will extend our study to one-variable continuous models, expressed in the form of ordinary differential equations. Once again, we will identify all possible dynamical regimes that can arise within this framework. In contrast to discrete-time models, we will discover that the range of behaviors is considerably more limited: In particular, one-dimensional continuous systems cannot exhibit chaotic dynamics.

Chapters 4 and 5 will be devoted to multivariable continuous and discrete-time models. We will see that chaotic behavior can occur only in continuous systems with three or more variables. Moreover, we will learn that chaotic regimes are closely linked to the existence of solutions with highly intricate geometric structure. These geometric objects, known as fractals, will be the focus of Chapter 6.

One of the central themes of this textbook is the transition between different dynamical regimes, a phenomenon associated with so-called bifurcations in parameter-dependent models. Although such transitions already appear in Chapter 2, a systematic treatment of bifurcation theory will be presented in Chapter 3, where we develop the necessary concepts in the context of continuous dynamical systems.

Finally, we note that continuous models expressed in terms of ordinary differential equations are fundamental throughout physics. Their prominence is not accidental: The entire theory of continuous dynamical systems has its roots in classical mechanics, where the motion of particles and rigid bodies is naturally described by differential equations. Newton's laws, in particular, provide direct relationships between forces and the resulting accelerations, making differential equations the natural tool for analysis of mechanical processes. To make this statement more evident, let us consider an example from mechanics.

Example 1.12. A particle is projected vertically upwards from ground level with speed v_0. Its speed, v, changes over time according to the equation

$$\dot{v} = -g - kv, \tag{1.9}$$

where g is the gravitational acceleration and the term kv, where $0 < k \ll 1$, accounts for air resistance. Find

(i) The time when the particle reaches its maximum height.
(ii) The time when it hits the ground.

Solution:
We start with finding the velocity $v(t)$ for what we integrate the equation (1.9). Using separation of variables we obtain

$$\int \frac{dv}{kv+g} = \int -dt \quad \Rightarrow \quad \ln\left(v + \frac{g}{k}\right) = -kt + c \quad \Rightarrow \quad v + \frac{g}{k} = Ae^{-kt},$$

where A is a constant of integration. Using the initial condition $v(0) = v_0$ we find

$$A = v_0 + \frac{g}{k} \quad \Rightarrow \quad v(t) = -\frac{g}{k} + \left(v_0 + \frac{g}{k}\right)e^{-kt}.$$

(i) The maximum height is reached when $v = 0$, and thus the time, t_1, when the particle is at the top of its trajectory is the root of the equation

$$0 = -\frac{g}{k} + \left(v_0 + \frac{g}{k}\right)e^{-kt},$$

hence

$$t_1 = \frac{1}{k}\ln\left(\frac{kv_0 + g}{g}\right) = \frac{1}{k}\ln\left(1 + \frac{kv_0}{g}\right).$$

As it is stated that $k \ll 1$, we can check the first-order approximation of this time. Using the second-order approximation of the logarithm, $\ln(1 + x) \approx x - x^2/2$, we obtain

$$t_1 \approx \frac{1}{k}\left(\frac{kv_0}{g} - \frac{k^2v_0^2}{2g^2}\right) = \frac{v_0}{g} - \frac{kv_0^2}{2g^2}.$$

(ii) Let $y(t)$ denote the position of the particle. Since $\dot{y} = v(t)$, we have

$$y(t) = \int_0^t v(\tau)\, d\tau = -\frac{g}{k}t + \frac{1}{k}\left(v_0 + \frac{g}{k}\right)(1 - e^{-kt}),$$

with $y(0) = 0$.
At the time, t_2, when the particle hits the ground, we have $y = 0$, and thus this time is a root of

$$-\frac{g}{k}t + \frac{1}{k}\left(v_0 + \frac{g}{k}\right)(1 - e^{-kt}) = 0. \tag{1.10}$$

This equation gives the required answer implicitly. To find the time t_2 when the particle hits the ground explicitly, we consider its first-order approximation. This is obtained by replacing the exponential term with its second-order Taylor approximation

$$e^{-kt} \approx 1 - kt + \frac{k^2t^2}{2}.$$

Hence,

$$\left(v_0 + \frac{g}{k}\right)(1 - e^{-kt}) \approx ktv_0 + gt - \frac{k^2t^2v_0}{2} - \frac{gkt^2}{2}.$$

Substituting this into equation (1.10) and simplifying yields

$$tv_0 - \frac{gt^2}{2} - \frac{kv_0t^2}{2} = 0.$$

This is a quadratic equation with two roots: $t = 0$ (the particle is initially at ground level), and

$$t_2 = \frac{2v_0}{g + v_0k} = \frac{2v_0}{g}\left(1 + \frac{v_0k}{g}\right)^{-1} \approx \frac{2v_0}{g}\left(1 - \frac{v_0k}{g}\right),$$

which gives the approximation of the time when the falling particle hits the ground.

1.5 Important formulas

1. **Numbers and numerical systems**
   Numbers are represented by sets of digits

$$a_1a_2a_3\dots a_n.b_1b_2\dots b_k\dots,$$

where the a_is represent the integer part and the b_is the fractional part, with $a_1 \neq 0$ if $n > 1$. Different numeral systems have different numbers of digits. In the decimal system, there are 10 digits: The a_is and b_is range from 0 to 9. In the binary system, there are only two digits, so the a_is and b_is can be either 0 or 1.
Rational numbers differ from irrational numbers by the structure of their fractional part. Rational numbers are either integers (where $b_i = 0$ for all i), terminating decimals (there exists an n such that $b_i = 0$ for all $i > n$), or repeating decimals (there exist n and l such that $b_i = b_{i+l}$ for all $i > n$). The latter are denoted as

$$a_1a_2a_3...a_n.b_1b_2...b_n(b_{n+1}b_{n+2}...b_{n+l}).$$

The fractional part of irrational numbers contains an infinite and non-repeating sequence of digits.

2. **Quadratic formula**
   The roots of quadratic equation $ax^2 + bx + c = 0$ are given by

$$x_{1,2} = \frac{-b \pm \sqrt{b^2 - 4ac}}{2a}.$$

3. **Composite functions and their differentiation**
   Consider the composite function $f^{(2)}(x) = f(f(x))$. The chain rule for its differentiation is

$$\left.\frac{df^{(2)}}{dx}\right|_{x=x_1} = \left.\frac{df}{dx}\right|_{x=f(x_1)} \left.\frac{df}{dx}\right|_{x=x_1}.$$

4. **Sum of geometric progression**
   The sum of the infinite geometric progression $a, ar, ar^2, \dots, ar^n, \dots$, where $|r| < 1$, is

$$a + ar + ar^2 + \cdots + ar^n + \cdots = \sum_{i=0}^{\infty} ar^i = \frac{a}{1-r}.$$

5. **Complex numbers**
   Complex numbers can be represented in two ways: $z = x + iy$ or $z = re^{i\theta}$, where $r = \sqrt{x^2 + y^2}$ is the modulus and $\theta = \arctan\frac{y}{x}$ is the argument of the complex number z.
6. **Linear differential equation**
   The solution of the linear differential equation

$$\dot{x} = ax + b,$$

   with the initial condition $x(t = 0) = x_0$, is

$$x(t) = \left(x_0 + \frac{b}{a}\right)e^{at} - \frac{b}{a}.$$

7. **Eigensolution**
   Matrix $A = \begin{pmatrix} a_{11} & a_{12} \\ a_{21} & a_{22} \end{pmatrix}$ is characterized by its determinant

$$\det A = a_{11}a_{22} - a_{21}a_{12},$$

   trace,

$$\operatorname{tr} A = a_{11} + a_{22},$$

   eigenvalues, $\lambda_{1,2}$, which are roots of

$$\begin{vmatrix} a_{11} - \lambda & a_{12} \\ a_{21} & a_{22} - \lambda \end{vmatrix} = 0,$$

or

$$\lambda_{1,2} = \frac{\operatorname{tr} A \pm \sqrt{\operatorname{tr}^2 A - 4 \det A}}{2},$$

and eigenvectors, $\boldsymbol{b}_i = (b_{i1}, b_{i2})$, which are solutions of the equation

$$(a_{11} - \lambda_i)b_{i1} + a_{12}b_{i2} = 0$$

(or $a_{21}b_{i1} + (a_{22} - \lambda_i)b_{i2}$).

8. **Important Taylor expansions**

$$e^x = \sum_{n=0}^{n=\infty} \frac{x^n}{n!},$$

$$\ln(1+x) = \sum_{n=1}^{n=\infty} (-1)^{n+1} \frac{x^n}{n},$$

$$(1+x)^\alpha = \sum_{n=0}^{n=\infty} \binom{\alpha}{n} x^n, \quad \text{where} \quad \binom{\alpha}{n} = \frac{\alpha(\alpha-1)\dots(\alpha-n+1)}{n!}.$$

9. **Approximations of functions of two variables**

Linear approximation of the function $f(x,y)$ at the point (a,b):

$$f(x,y) \approx f(a,b) + f'_x(a,b)\,(x-a) + f'_y(a,b)\,(y-b),$$

and its quadratic approximation:

$$f(x,y) \approx f(a,b) + f'_x(a,b)(x-a) + f'_y(a,b)(y-b) + \frac{1}{2} f''_{xx}(a,b)(x-a)^2$$
$$+ f''_{xy}(a,b)(x-a)(y-b) + \frac{1}{2} f''_{yy}(a,b)(y-b)^2,$$

where $|x-a| \ll 1$ and $|y-b| \ll 1$.

1.6 Exercises

These exercises are for the revision of basic mathematical techniques that will be used in the following chapters of the textbook.

Problem 1.1. The function $f(x) = 1 - x^2$ is defined on the domain $0 \le x \le 1$.

(i) Find $F(x) = f(f(x))$ and evaluate dF/dx.

(ii) Find the inverse function $f^{-1}(x)$.

Problem 1.2. Given that the polynomial equation

$$-x^3 - 2x^2 + 2x + 1 = 0$$

has one root $x = 1$, find the other two roots.

Problem 1.3. If $-1 < 3 - 2\mu < 1$, find the region of μ allowed.

Problem 1.4. Find d such that $4^n/(\frac{1}{3^n})^{-d} \rightarrow C$, when $n \rightarrow \infty$, where C is a nonzero constant.

Problem 1.5. Show that

$$\ln(e+1) = 1 + \sum_{n=1}^{\infty}(-1)^{n+1}\frac{e^{-n}}{n}.$$

How many terms from the sum on the right-hand side are needed to evaluate this number to the second decimal place?

Problem 1.6. Given the differentiable function of two variables $f(x,y) = x^2 + xy$, find its partial derivatives $\frac{\partial f}{\partial x}$ and $\frac{\partial f}{\partial y}$.

Deduce a linear approximation to $f(x,y)$ near the point $(x,y) = (1,2)$.

Problem 1.7. Find the eigenvalues of the matrix

$$A = \begin{pmatrix} a & -1 \\ 1 & 0 \end{pmatrix},$$

where a is real, distinguishing between the cases $a^2 > 4$ and $a^2 < 4$.

Problem 1.8. Solve the differential equation

$$\dot{y} = -ay(y-1),$$

where a is a real positive constant and $y(0) = 2$.

1.7 Solutions to exercises

Solution to Problem 1.1. $f(x) = 1 - x^2$

(i) Composite function

$$F(x) = f(f(x)) = 1 - f^2(x) = 1 - (1-x^2)^2.$$

Derivative of the composite function

$$\frac{dF}{dx} = \frac{dF}{df}\frac{df}{dx} = (-2f(x))(-2x) = -2(1-x^2)(-2x) = 4x(1-x^2).$$

(ii) The function $f(x) = 1 - x^2$ can be rewritten as $x^2 = 1 - f(x)$.
Since $x \in [0,1]$, we choose the positive root $x = \sqrt{1-f(x)}$.
Updating notation, the inverse function is $f^{-1}(x) = \sqrt{1-x}$.

Solution to Problem 1.2. Using the known root, we can factorize the cubic polynomial as follows:

$$-x^3 - 2x^2 + 2x + 1 = (x - 1)(-x^2 + Ax - 1),$$

where the unknown coefficient A is determined by matching the coefficients of the x and x^2 terms on both sides of the equation.

For the x^2 terms we obtain $-2 = 1 + A$, which gives $A = -3$.

For the x terms, $2 = -1 - A$, which again gives $A = -3$.

Thus, the two remaining roots of the original cubic polynomial are the roots of the quadratic equation

$$-x^2 - 3x - 1 = 0,$$

which are

$$x_{\pm} = \frac{-3 \pm \sqrt{5}}{2}.$$

Solution to Problem 1.3. Starting from

$$-1 < 3 - 2\mu < 1,$$

we multiply all parts of the inequality by -1, which reverses the inequality signs

$$-1 < 2\mu - 3 < 1.$$

Next, we add 3 to all parts and then divide the entire inequality by 2 to obtain

$$1 < \mu < 2.$$

Solution to Problem 1.4. To find the value of d such that

$$4^n / \left(\frac{1}{3^n}\right)^{-d} \to C, \quad \text{as } n \to \infty,$$

where C is a no-zero constant, we observe that

$$4^n / \left(\frac{1}{3^n}\right)^{-d} = \left(\frac{4}{3^d}\right)^n.$$

As $n \to \infty$, this expression tends to 0 if $4 < 3^d$ and diverges to infinity if $4 > 3^d$. For the expression to converge to a nonzero constant C, we must have $4 = 3^d$, which leads to

$$d = \frac{\log 4}{\log 3}.$$

Solution to Problem 1.5.

$$\ln(e+1) = \ln(e(1+e^{-1})) = \ln e + \ln(1+e^{-1}) = 1 + \sum_{n=1}^{\infty}(-1)^{n+1}\frac{e^{-n}}{n}.$$

Here we used the Taylor expansion of the logarithm

$$\ln(1+x) = x - \frac{x^2}{2} + \frac{x^3}{3} - \cdots = \sum_{n=1}^{\infty}(-1)^{n+1}\frac{x^n}{n}.$$

To evaluate $\ln(e+1)$ to two decimal places, we sum all terms whose absolute values exceed 0.001 since smaller terms do not affect rounding at that precision. The fifth term, $e^{-5}/5$, is still greater than 0.001, while the sixth term is already smaller and can be ignored.

Adding the first four terms of the series to 1, gives approximately 1.31. Including the fifth term does not change the value at the second decimal place. Therefore,

$$\ln(e+1) \approx 1.31.$$

Solution to Problem 1.6. Partial derivatives $\frac{\partial f}{\partial x}$ and $\frac{\partial f}{\partial y}$ are given by

$$\frac{\partial f}{\partial x} = 2x + y, \quad \frac{\partial f}{\partial y} = x.$$

A linear approximation to $f(x,y)$ near the point $(1,2)$ is provided by the first-order Taylor expansion

$$f(x,y) = f(1,2) + \left.\frac{\partial f}{\partial x}\right|_{(1,2)}(x-1) + \left.\frac{\partial f}{\partial y}\right|_{(1,2)}(y-2) = 3 + 4(x-1) + (y-2).$$

Solution to Problem 1.7. Eigenvalues are the roots of the characteristic polynomial

$$\begin{vmatrix} a-\lambda & -1 \\ 1 & -\lambda \end{vmatrix} = (a-\lambda)(-\lambda) - (-1) = \lambda^2 - a\lambda + 1,$$

which are

$$\lambda_{1,2} = \frac{a \pm \sqrt{a^2-4}}{2}.$$

For $a^2 > 4$, the eigenvalues are real. For $a^2 < 4$, they form a complex conjugate pair:

$$\lambda_{1,2} = \frac{a \pm i\sqrt{4-a^2}}{2}.$$

Solution to Problem 1.8. We separate the variables and integrate

$$\int \frac{dy}{y(y-1)} = -a\int dt.$$

Using partial fractions this becomes

$$\int\left(\frac{1}{y-1}-\frac{1}{y}\right)dy=-at+c,$$

where c is an integration constant.

Since $y\geq 2$, both y and $y-1$ are positive, so we obtain

$$\log\left(\frac{y-1}{y}\right)=-at+c,$$

or equivalently

$$\frac{y-1}{y}=Ke^{-at},\quad K=e^{c}.$$

Solving for y gives

$$y=\frac{1}{1-Ke^{-at}}.$$

The constant K is determined from the initial condition:

$$y(0)=2=\frac{1}{1-K}\quad\Rightarrow\quad K=0.5.$$

Therefore, the solution is

$$y(t)=\frac{2}{2-e^{-at}}.$$

2 One-variable discrete-time dynamical systems

One-variable discrete-time models are represented by the difference equation

$$u_{n+1} = f(u_n, n),$$

which is commonly referred to as a one-dimensional (or 1D) map. The solution of such an equation is represented by a set of numbers: $u_0, u_1, \ldots, u_n, \ldots$, and is called an *orbit*. It is uniquely defined by the map-generating function $f(u, n)$ and the initial value u_0. For a so-called autonomous map, the map-generating function does not explicitly depend on the time step n

$$u_{n+1} = f(u_n).$$

A special case is a *linear map*

$$u_{n+1} = a(n)u_n + b(n),$$

which, in the case of an autonomous system, has constant coefficients

$$u_{n+1} = au_n + b. \tag{2.1}$$

Example 2.1. Find the monthly repayment for a loan of £10,000, taken at an interest rate of 6.1 % per annum and to be repaid over 5 years (60 months).

Solution:
Let u_n be the size of the debt in the nth month. Our goal is to determine how it depends on u_{n-1}, i. e., to find $f(u)$ such that $u_n = f(u_{n-1})$. According to the problem's conditions: $u_0 = 10{,}000$, and also $u_{60} = 0$—the debt is paid off in 5 years or 60 months. Let us introduce two parameters: α is the monthly interest rate, and β is the size of the monthly payment, which we have to find. Then, the change in the balance (debt) over a one-month period is given by the difference equation

$$u_n = u_{n-1} + \alpha u_{n-1} - \beta = (1+\alpha)u_{n-1} - \beta,$$

where the term αu_{n-1} represents the increase in balance due to the interest charged and β its reduction due to the monthly payment.

Finding α. In the absence of payments ($\beta = 0$), the balance after the first month is $u_1 = (1+\alpha)u_0$, after two months, $u_2 = (1+\alpha)u_1 = (1+\alpha)^2 u_0$; and after 1 year (i. e., 12 months), $u_{12} = u_0(1+\alpha)^{12}$. This should be equivalent to the annual (12-month) interest rate

$$u_{12} = u_0\left(1 + \frac{6.1}{100}\right) = 1.061u_0.$$

https://doi.org/10.1515/9783111348339-002

Thus,

$$(1+\alpha)^{12} = 1.061 \quad \Rightarrow \quad \alpha = \sqrt[12]{1.061} - 1 = 0.00495.$$

This answer can be estimated more simply using the first-order approximation, $(1+x)^n \approx 1+nx$, which comes from the Taylor expansion and is valid when $|x| \ll 1$. Applying this to the formula for α above, we have

$$\alpha = \sqrt[12]{1+0.061} - 1 \approx 1 + \frac{0.061}{12} - 1 = \frac{0.061}{12} = 0.00508,$$

and this approximation is not too different from the more accurate value found before. Now that we know α, we can find β. This can be done in two ways.

1st way (involves direct calculations).

$$\begin{aligned}
u_1 &= (1+\alpha)u_0 - \beta, \\
u_2 &= (1+\alpha)u_1 - \beta = (1+\alpha)[(1+\alpha)u_0 - \beta] - \beta \\
&= (1+\alpha)^2 u_0 - \beta(1+\alpha) - \beta, \\
u_3 &= (1+\alpha)u_2 - \beta = \cdots \\
&= (1+\alpha)^3 u_0 - \beta(1+\alpha)^2 - \beta(1+\alpha) - \beta, \\
&\vdots \\
u_n &= (1+\alpha)^n u_0 - \beta[1 + (1+\alpha) + \cdots + (1+\alpha)^{n-1}].
\end{aligned}$$

The expression in the square brackets is the sum of a geometric progression—with common ratio $q = 1+\alpha$. Using the formula for the sum of the geometric progression $1 + q + q^2 + \cdots + q^{n-1} = (1-q^n)/(1-q)$, we get

$$\begin{aligned}
u_n &= (1+\alpha)^n u_0 - \beta[1 + (1+\alpha) + \cdots + (1+\alpha)^{n-1}] \\
&= (1+\alpha)^n u_0 - \beta \frac{1-(1+\alpha)^n}{1-(1+\alpha)} \\
&= (1+\alpha)^n u_0 + \frac{\beta}{\alpha}(1-(1+\alpha)^n) = \left(u_0 - \frac{\beta}{\alpha}\right)(1+\alpha)^n + \frac{\beta}{\alpha}.
\end{aligned}$$

2nd way (using substitution). In the original difference equation

$$u_n = (1+\alpha)u_{n-1} - \beta,$$

and subtracting β/α from both sides yields

$$\begin{aligned}
u_n - \frac{\beta}{\alpha} = (1+\alpha)u_{n-1} - \beta - \frac{\beta}{\alpha} &= (1+\alpha)u_{n-1} - \left(\frac{\beta}{\alpha}\right)(1+\alpha) \\
&= (1+\alpha)\left(u_{n-1} - \frac{\beta}{\alpha}\right).
\end{aligned}$$

After the substitution $x_n = u_n - \beta/\alpha$, the formula transforms into

$$x_n = (1+\alpha)x_{n-1}, \quad \text{where } x_n = u_n - \frac{\beta}{\alpha} \text{ and } x_{n-1} = u_{n-1} - \frac{\beta}{\alpha}.$$

So,

$$x_1 = (1+\alpha)x_0, \quad x_2 = (1+\alpha)x_1 = (1+\alpha)^2 x_0, \ldots,$$

leading to the formula for x_n in terms of x_0

$$x_n = (1+\alpha)^n x_0.$$

Using backward substitution, $u_n = x_n + \beta/\alpha$ gives again

$$u_n = \left(u_0 - \frac{\beta}{\alpha}\right)(1+\alpha)^n + \frac{\beta}{\alpha}.$$

Using the conditions $u_0 = 10^4$ and $u_{60} = 0$, we get the equation for β

$$\begin{aligned} U_{60} = 0 &= \left(10000 - \frac{\beta}{\alpha}\right)(1+\alpha)^{60} + \frac{\beta}{\alpha} \quad \Rightarrow \\ \frac{\beta}{\alpha} &= \left(\frac{\beta}{\alpha} - 10000\right)(1+\alpha)^{60} \quad \Rightarrow \\ \beta &= \alpha \frac{10000}{1-(1+\alpha)^{-60}}, \end{aligned}$$

where $\alpha = 0.00495$, and thus $\beta \approx$ £193.05. Thus, the monthly repayment for a loan of £10,000, taken for five years at an interest rate of 6.1 % per annum, is £193.05.

2.1 Fixed points

The solution of the autonomous linear map (2.1) is

$$u_n = \left(u_0 - \frac{b}{1-a}\right)a^n + \frac{b}{1-a}.$$

This can be derived using one of the two methods explained earlier. The solution consists of two terms: The first depends on the iteration number and is called the *transient solution*, while the second does not depend on n and is therefore called the *stationary solution*. Indeed, if $u_0 = b/(1-a)$, then the transient solution is zero, and thus $u_n = u_0 \ \forall n$.

Definition 2.1. Consider the map $u_{n+1} = f(u_n)$. Solutions such that $u_n = u_0$ for all n are called **stationary solutions**. Furthermore, values U such that, if $u_0 = U$, it follows that $u_n = U$ for all n, are called **fixed points** (also known as *stationary or equilibrium points*).

From $u_n = U$, it follows that $u_{n+1} = f(u_n) = U$, and thus fixed points are the roots of the algebraic equation

$$u = f(u).$$

Applied to the linear map (2.1), this gives $u = au + b$, and therefore the fixed point of this map is

$$U = \frac{b}{1-a}.$$

Obviously, this formula can be used only when $a \neq 1$. If $a = 1$, then the fixed point is a root of $u = u + b$, which has no solutions if $b \neq 0$ and is an identity (valid for any value of u) if $b = 0$. In Example 2.1 just considered, we used slightly different notations, i. e., $a = 1 + \alpha$ and $b = -\beta$, and the fixed point for that example was

$$U = \frac{b}{1-a} = \frac{-\beta}{1-(1+\alpha)} = \frac{\beta}{\alpha}.$$

For either of these linear maps, the solution can be written as

$$u_n = (u_0 - U)a^n + U.$$

The second (substitution) method for solving Example 2.1 shows that it is easier to find the solution in terms of the difference between u_n and the fixed point U. That is, if $x_n = u_n - U$, then the equation for x_n no longer contains the second (free) term $x_n = ax_{n-1}$. The solution in terms of x is $x_n = a^n x_0$, and after back-substitution, we obtain $u_n = a^n(u_0 - U) + U$. The equation $x_n = ax_{n-1}$ is called *homogeneous* (i. e., it contains no free terms). As we have seen, its solution is $x_n = a^n x_0$.

The solution of a one-dimensional map can be constructed or illustrated using *cobwebbing*. This is shown in Figure 2.1, which also demonstrates that a fixed point corresponds to the intersection between the diagonal line $u_{n+1} = u_n$ and the graph of the function $u_{n+1} = f(u_n)$ on the cobweb diagram.

2.2 Second-order maps

The result can be extended to the solutions of second-order homogeneous linear equations $x_n = ax_{n-1} + bx_{n-2}$. While the solution of the first-order equation is given by a geometric series, the solution of the second-order equation is given by a linear combination of two geometric series

$$x_n = \lambda_1^n C_1 + \lambda_2^n C_2,$$

where λ_1 and λ_2 are the roots of the characteristic equation $\lambda^2 = a\lambda + b$, and the constants C_1 and C_2 are determined from the initial conditions, that is, from the values of u_0 and u_1.

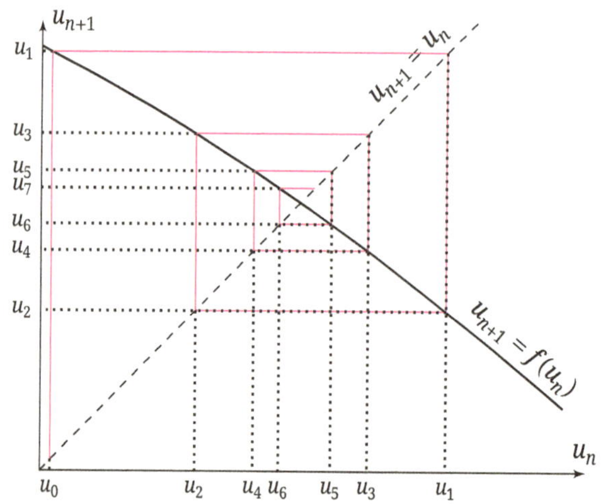

Figure 2.1: Cobweb diagram. The solid line shows the plot of $u_{n+1} = f(u_n)$, the dashed line is the diagonal $u_{n+1} = u_n$, and the red line represents the graphical method for constructing the solution (cobwebbing). The vertical segments of the red line have horizontal coordinate u_n, while the horizontal segments have vertical coordinate u_{n+1}.

Example 2.2. The famous Fibonacci sequence $(0, 1, 1, 2, 3, 5, 8, 13, \ldots,$ where each number is the sum of the two preceding ones) turns out to be the solution of the lesser-known Fibonacci model. This model, proposed in the 12th century, describes the dynamics of a rabbit population and is based on the following assumptions:

1. Rabbits produce offspring once per year.
2. Each litter consists of one pair of rabbits.
3. Rabbits live for two years.

How does the size of the population change over time, if we start with a single pair of rabbits?

Solution:

Let us look at the dynamics of the population, following the rules given above and assuming that the original pair of rabbits is in their first year.

At $t = 0$ (initially), there is one pair of newborn rabbits.

At $t = 1$, there are two pairs; the original pair, which is now one-year old, produces one newborn pair.

At $t = 2$, there are 3 pairs: 2 newborn pairs and one one-year-old pair, while the original pair becomes two-year-old and dies.

And so on.

Let u_n be the number of rabbit pairs in year n, consisting of two subgroups: young pairs (aged 0) and old pairs (aged 1). Then,

$$
\begin{aligned}
t = 0 &: u_0 = u_0[\text{young}] \\
t = 1 &: u_1 = u_0[\text{young}] + u_0[\text{old}] \\
t = 2 &: u_2 = u_1[\text{young}] + u_0[\text{old}] \\
t = 3 &: u_3 = u_2[\text{young}] + u_1[\text{old}] \\
&\vdots
\end{aligned}
$$

We can conclude (and prove, for example, using the method of mathematical induction) that

$$u_n = u_{n-1} + u_{n-2}. \tag{2.2}$$

This formula is known to produce the Fibonacci sequence $(0, 1, 1, 2, 3, 5, \ldots)$, and the solution to our problem is given by this sequence, but with the first two terms omitted, i. e., $1, 2, 3, 5, \ldots$. It is also known that the ratio of consecutive numbers in the Fibonacci sequence, u_n/u_{n-1}, tends to a limit as $n \to \infty$, and this limit is the number

$$\lim_{n\to\infty} \frac{u_n}{u_{n-1}} = \frac{\sqrt{5}+1}{2}, \tag{2.3}$$

which is called the *golden ratio*. It is easy to see that

$$\frac{\sqrt{5}+1}{2} = \frac{2}{\sqrt{5}-1}.$$

Formula (2.3) can be proved as follows. The model (2.2) is a linear, homogeneous, second-order difference equation. Its solution is given by

$$u_n = C_1\lambda_1^n + C_2\lambda_2^n, \tag{2.4}$$

where $\lambda_{1,2}$ are roots of the characteristic equation

$$\lambda^2 - \lambda - 1 = 0,$$

and C_1 and C_2 are arbitrary constants. This characteristic equation is derived by assuming a solution of the form $u_n = \lambda^n$. Substituting into $u_n = u_{n-1} + u_{n-2}$ gives

$$\lambda^n = \lambda^{n-1} + \lambda^{n-2},$$

which simplifies to

$$\lambda^2 = \lambda + 1.$$

The roots of this equation are

$$\lambda_{1,2} = \frac{1}{2} \pm \sqrt{\frac{1}{4} + 1} = \frac{1 \pm \sqrt{5}}{2}.$$

Thus, the solutions of (2.2) are given by

$$u_n = C_1\lambda_1^n + C_2\lambda_2^n = C_1\left(\frac{1+\sqrt{5}}{2}\right)^n + C_2\left(\frac{1-\sqrt{5}}{2}\right)^n,$$

where C_1 and C_2 are constants, whose values are determined by the initial conditions (that is, the values of u_0 and u_1). Recall that

$$\lim_{n\to\infty} |a|^n = \begin{cases} \infty, & \text{if } |a| > 1, \\ 0, & \text{if } |a| < 1, \\ 1, & \text{if } |a| = 1. \end{cases}$$

So, since

$$\lambda_1 = \frac{1-\sqrt{5}}{2} = -0.618 \quad (\text{i. e., } -1 < \lambda_1 < 0)$$

and

$$\lambda_2 = \frac{1+\sqrt{5}}{2} = 1.618 > 1,$$

we conclude that the term involving λ_1^n vanishes as $n \to \infty$, and thus,

$$u_n \to C_1\left(\frac{1+\sqrt{5}}{2}\right)^n \quad \text{as } n \to \infty.$$

Therefore,

$$\lim_{n\to\infty} \frac{u_n}{u_{n-1}} = \frac{C_1(\frac{1+\sqrt{5}}{2})^n}{C_1(\frac{1+\sqrt{5}}{2})^{n-1}} = \frac{1+\sqrt{5}}{2}.$$

From Eq. (2.2) we see that, in order to compute u_n with $n \geq 2$, we must first specify the values of u_0 and u_1. This initial values are then used to determine the constants C_1 and C_2 in the general solution (2.4). For example, if $u_0 = 0$ and $u_1 = 1$, we can determine the constants C_1 and C_2 as follows.

Substituting $n = 0$ into the general solution (2.4) yields

$$u_0 = C_1\left(\frac{1+\sqrt{5}}{2}\right)^0 + C_2\left(\frac{1-\sqrt{5}}{2}\right)^0 = C_1 + C_2 = 0,$$

which implies $C_1 = -C_2$.

Substituting $n = 1$,

$$u_1 = C_1\left(\frac{1+\sqrt{5}}{2}\right)^1 + C_2\left(\frac{1-\sqrt{5}}{2}\right)^1.$$

Substituting $C_1 = -C_2$ in the previous expression,

$$u_1 = C_1\left(\frac{1+\sqrt{5}}{2}\right) - C_1\left(\frac{1-\sqrt{5}}{2}\right) = C_1\frac{1+\sqrt{5}-(1-\sqrt{5})}{2} = \sqrt{5}C_1 = 1.$$

Therefore,

$$C_1 = \frac{1}{\sqrt{5}}, \quad C_2 = -\frac{1}{\sqrt{5}}.$$

Thus, any number in the Fibonacci sequence $0, 1, 1, 2, 3, 5, \dots$ can be expressed using the following closed-form formula:

$$u_n = \frac{1}{\sqrt{5}}\left[\left(\frac{1+\sqrt{5}}{2}\right)^n - \left(\frac{1-\sqrt{5}}{2}\right)^n\right].$$

This expression is known as *Binet's formula*. One can easily verify its correctness for $n = 2$ in

$$u_2 = \frac{1}{\sqrt{5}}\left[\left(\frac{1+\sqrt{5}}{2}\right)^2 - \left(\frac{1-\sqrt{5}}{2}\right)^2\right] = \cdots = 1.$$

2.3 Stability of fixed points

As we have seen before, the solution of the linear map $u_{n+1} = au_n + b$ is $u_n = (u_0 - U)a^n + U$, where $U = \frac{b}{1-a}$ is the fixed point of this map. The first term in this expression (the so-called *transient solution*) gives the distance between u_n and the fixed point U. It is easy to see that if $|a| < 1$, this term tends to zero as $n \to \infty$, and therefore $u_n \to U$, meaning the orbit converges to the fixed point. If $|a| > 1$, then the transient part of the solution grows, and the orbit diverges from the fixed point. Thus, an important property of orbits is their relationship with fixed points: Orbits can either converge to or diverge from fixed points.

Definition 2.2. Consider a fixed point U of the map $u_{n+1} = f(u_n)$. If there exists δ such that, from $|u_0 - U| < \delta$, it follows that $u_n \to U$ as $n \to \infty$, then the fixed point U is called **asymptotically stable**.

This definition refers to the solution converging to the fixed point. There is a weaker definition of stability, which indicates that the solution does not diverge from the fixed point.

Definition 2.3. Consider a fixed point U of the map $u_{n+1} = f(u_n)$. If, for any ε, there exists δ such that from $|u_0 - U| < \delta$, it follows that $|u_0 - U| < \varepsilon$, for any n, and then the fixed point U is called **stable** or **neutrally stable**. Otherwise, the fixed point is called **unstable**.

The nature of the fixed point is determined by the characteristics of the map-generating function $f(u)$, as stated in the following theorem.

Theorem 2.1. *A fixed point, U, of the map $u_{n+1} = f(u_n)$ is asymptotically stable if* $|f'(u = U)| < 1$.

Proof. Replace the orbit $u_0, u_1, \ldots, u_n, \ldots$ with the sequence $\varepsilon_0, \varepsilon_1, \ldots, \varepsilon_n, \ldots$, representing the distance between the orbit and the fixed point U: $u_n = U + \varepsilon_n$. Assume that $|\varepsilon_n| \ll 1$, and replace the right-hand side of the equation $u_{n+1} = f(u_n)$ with its first-order approximation

$$u_{n+1} = U + \varepsilon_{n+1} = f(u_n) = f(U + \varepsilon_n) \approx f(U) + f'(U)\varepsilon_n.$$

As $U = f(U)$ (since U is a fixed point), this equation simplifies, and the first-order approximation to solutions near the fixed point is given by

$$\varepsilon_{n+1} = f'(U)\,\varepsilon_n,$$

for which the solution is

$$\varepsilon_n = \left[f'(U)\right]^n \varepsilon_0.$$

This solution tends to zero (and therefore the original orbit u_n tends to U) when $|f'(U)| < 1$. The proof is complete. □

Thus, while the fixed points can be found graphically as points of intersection between the map-generating function $f(x)$ and the diagonal line, $y = x$, their stability is determined by the slope of the function $f(x)$ at these points. At a stable fixed point X, the function $f(x)$ crosses the line, $y = x$, 'from above to below'—that is, it is either increasing more slowly or decreasing (but not too steeply), so that at this point $-1 < f'(x = X) < 1$.

This is illustrated in Figure 2.2, where a map-generating function crossing the diagonal line at three points is shown. The corresponding map has three fixed points: $X_1 = 0$, $X_2 = 0.5$, and $X_3 = 1$, corresponding to the three points of intersection between the function $f(x)$ and the diagonal line $y = x$. The fixed points $X_1 = 0$ and $X_3 = 1$ are stable since the slope of the function $f(x)$ is positive and less than 1 at these points (i. e., the function $f(x)$ increases more slowly than the diagonal line, whose slope is 1). The fixed point at $x = 0.5$ is unstable since the slope of $f(x)$ at this point is greater than 1 (i. e., it increases faster than the diagonal line).

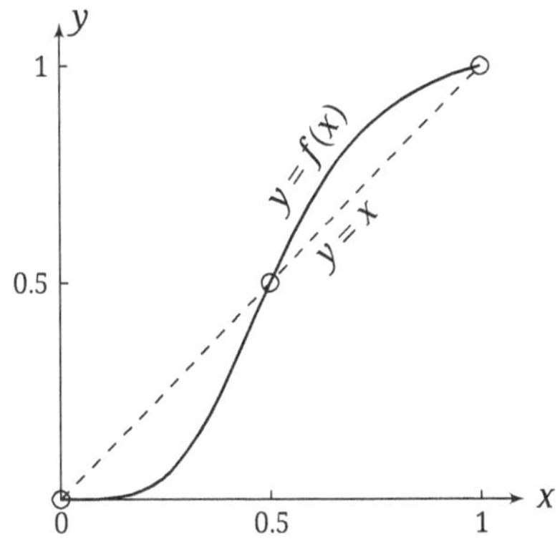

Figure 2.2: A plot of the map-generating function $f(x)$ with $x \in [0,1]$ is shown by a solid line. It intersects the diagonal line $y = x$ (shown by a dashed line) at three points: $X_1 = 0$, $X_2 = 0.5$, and $X_3 = 1$.

Definition 2.4. Stable fixed points are often called **attractors**. The set of all initial values u_0 for orbits that converge to the attractor is called the **basin of attraction** for that attractor. For example, for the function shown in Figure 2.2, orbits with $x_0 < 0.5$ converge to $x = 0$, while those with $x_0 > 0.5$ converge to $x = 1$. Thus, the basin of attraction of $X_1 = 0$ is $[0, 0.5)$, and the basin of attraction of $X_3 = 1$ is $(0.5, 1]$.

Example 2.3. The *tent map* is defined on the domain $u \in [0,1]$ as

$$u_{n+1} = \begin{cases} 2u_n, & \text{if } 0 \le u_n \le \frac{1}{2} \\ 2(1-u_n), & \text{if } \frac{1}{2} < u_n \le 1 \end{cases} \tag{2.5}$$

and is shown in Figure 2.3. Find the fixed points of this map and analyze their stability.

Solution:
As we see from Figure 2.3, there are two points of intersection between the map-generating function and the diagonal line. One of them lies on the first branch of the tent map, where $u_{n+1} = 2u_n$, intersecting the diagonal line $u_{n+1} = u_n$

$$\begin{cases} u_{n+1} = 2u_n, \\ u_{n+1} = u_n. \end{cases}$$

The solution of this system $u_{n+1} = u_n = 0$ gives the first fixed point $U_1 = 0$. The second fixed point corresponds to the point of intersection between the second branch of the tent map $u_{n+1} = 2(1-u_n)$ and the diagonal line $u_{n+1} = u_n$. It is found by solving the system

$$\begin{cases} u_{n+1} = 2(1-u_n), \\ u_{n+1} = u_n. \end{cases}$$

Substituting, we get

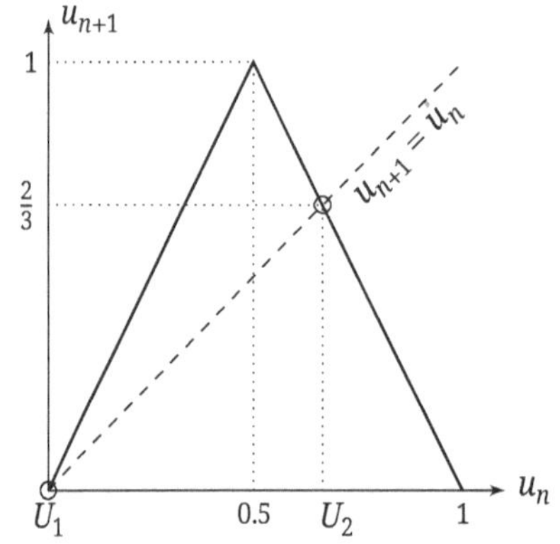

Figure 2.3: Tent map. The solid line represents the graph of the tent map defined by Eq. (2.5). The fixed points correspond to the intersections of this graph with the diagonal line $u_{n+1} = u_n$ (shown dashed), namely, $U_1 = 0$ and $U_2 = \frac{2}{3}$.

$$u_n = 2(1 - u_n) \quad \Rightarrow \quad U_2 = 2(1 - U_2) \quad \Rightarrow \quad 3U_2 = 2 \quad \Rightarrow \quad U_2 = \frac{2}{3}.$$

Now that we have found both fixed points, we can analyze their stability.

For the first fixed point, $U_1 = 0$ the map is defined by the first branch

$$f(u) = 2u \quad \Rightarrow \quad f'(u = U_1) = 2.$$

Since $f'(u = U_1) = 2 > 1$, the fixed point $U_1 = 0.$ is *unstable*, according to the stability criterion from the previous theorem.

For the second fixed point,

$$U_2 = \frac{2}{3}; \quad f(u) = 2(1 - u); \quad f'(u = U_2) = -2.$$

We observe that $|f'(U_2)| = 2 > 1$, and therefore $U_2 = 2/3$ is also *unstable*.

2.4 Logistic map

The precursor of the logistic map is the linear map

$$P_{n+1} = rP_n,$$

which represents a discrete-time analogue of the Malthusian model for the dynamics of human population. Here, P_n denotes the size of the population at time $n\Delta t$, or simply in year n, if the time step Δt is set to one year. The solution of the equation is given by

$$P_n = P_0 r^n,$$

which predicts exponential growth of the human population. This can be expressed as

$$P_n = P_0 r^n = P_0 (e^{\ln(r)})^n = P_0 e^{\ln(r)n}$$

which in turn can be rewritten in continuous-time form as

$$P(t) = P_0 e^{at},$$

where $a = \ln(r)$ defines the population growth rate (with $a > 0$ if $r > 1$) and $t = 0, 1, 2, \ldots, n, \ldots$ denotes time measured either in years—as in the original Malthusian model—or in units such as generations, depending on the context of the population dynamics.

The **logistic map**

$$u_{n+1} = r u_n (1 - u_n), \tag{2.6}$$

represents a modification of the linear growth model discussed earlier. Unlike the exponential model, this map introduces saturation, capturing the effect of limited resources on population growth. It is a discrete-time analogue of *the logistic equation*, which will be considered in the next chapter.

The logistic map is defined on the domain $u \in [0, 1]$, where $r > 0$ is a parameter defining the growth rate of the population. For initial conditions $u_0 \in [0, 1]$, the next iterate

$$u_1 = f(u_0) = r u_0 (1 - u_0)$$

lies is in the range $[0, r/4]$ since the function attains its maximum at $u = 1/2$, where $f(1/2) = r/4$. To ensure that all values u_n remain within the interval $[0, 1]$, it is necessary to impose the condition

$$0 < r \leq 4.$$

The graph of the logistic map is shown in Figure 2.4.

Let's find the fixed points of the logistic map (2.6) and determine their stability. Fixed points are values U such that $u_n = u_{n+1} = U$. Substituting into the logistic map, we get the equation

$$U = rU(1 - U),$$

which is a quadratic equation. This yields two solutions. The first fixed point is $U_1 = 0$, and the second satisfies the equation

$$1 = r(1 - U) \quad \Rightarrow \quad \frac{1}{r} = 1 - U \quad \Rightarrow \quad U_2 = 1 - \frac{1}{r}.$$

Hence, the two fixed points of the logistic map are

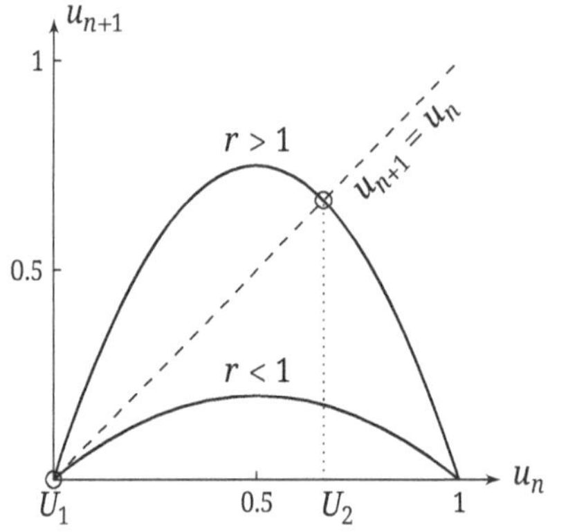

Figure 2.4: Logistic map. Plots of the logistic map for two values of the parameter r ($r = 0.8 < 1$ and $r = 3 > 1$) are shown. Fixed points are identified as the intersections of the graph of the map-generating function with the diagonal line $u_{n+1} = u_n$. When $r < 1$, the logistic map has only one fixed point, $U_1 = 0$. When $r > 1$, an additional fixed point appears at $U_2 = 1 - \frac{1}{r}$.

$$U_1 = 0, \quad U_2 = 1 - \frac{1}{r}. \tag{2.7}$$

Note that the second fixed point satisfies the following conditions:

$$\begin{cases} U_2 > 0, & \text{if } r > 1 \\ U_2 < 0, & \text{if } r < 1. \end{cases}$$

Since the logistic map is defined on the domain $u \in [0, 1]$, only fixed points within this domain are meaningful. Therefore, when $r < 1$, the map has a single fixed point at $U_1 = 0$. When $r > 1$, it has two fixed points $U_1 = 0$ and $U_2 = 1 - \frac{1}{r}$ (see Figure 2.4).

Let us determine the stability of these fixed points. For the logistic map $f(u) = ru(1-u)$, we compute the derivative

$$f'(u) = r - 2ru.$$

Evaluating it at the first fixed point $U_1 = 0$, we get

$$f'(U_1) = f'(0) = r.$$

Therefore, the fixed point U_1 is *stable* if $r < 1$ and *unstable* if $r > 1$.

Evaluating the derivative at the second fixed point $U_2 = 1 - \frac{1}{r}$, we get

$$f'(U_2) = r - 2r\left(1 - \frac{1}{r}\right) = r - 2r + 2 = 2 - r.$$

Hence, the fixed point U_2 is *stable* if $|2 - r| < 1$, which gives

$$-1 < 2 - r < 1 \quad \Rightarrow \quad 1 < r < 3.$$

Note, that when $r < 1$, the second fixed point U_2 does not exist, while when $r > 3$, it exists and *is unstable*.

Definition 2.5. A fixed point U is called **superstable** if $f'(U) = 0$.

This concept refers to the rapid (nearly instantaneous) convergence of nearby orbits to the fixed point. In terms of the distance from the fixed point, let $\varepsilon_n = u_n - U$. Then, from the linear approximation, we have

$$\varepsilon_1 = f'(U)\varepsilon_0.$$

If $f'(U) = 0$, it follows that $\varepsilon_1 = 0$ for any initial deviation ε_0. Hence, $u_1 = U$, and the orbit reaches the fixed point in a single step—indicating instantaneous convergence, at least in the linear approximation.

Getting back to the logistic map, we note that the fixed point $U_2 = 1 - 1/r$ is superstable when $f'(u = U_2) = 2 - r = 0$ or equivalently when $r = 2$. The solution of logistic map with $r = 2$

$$u_{n+1} = 2u_n(1 - u_n)$$

can be found analytically. Note that $U_2 = 1 - 1/r = 0.5$ when $r = 2$. Let's use the substitution $\varepsilon_n = u_n - U_2 = u_n - 1/2$

$$\begin{aligned} u_{n+1} = \varepsilon_{n+1} + \frac{1}{2} &= 2u_n(1 - u_n) = 2\left(\varepsilon_n + \frac{1}{2}\right)\left(1 - \frac{1}{2} - \varepsilon_n\right) \\ &= 2\left(\varepsilon_n + \frac{1}{2}\right)\left(\frac{1}{2} - \varepsilon_n\right) = 2\left(\frac{1}{4} - \varepsilon_n^2\right) = \frac{1}{2} - 2\varepsilon_n^2. \end{aligned}$$

So,

$$\varepsilon_{n+1} + \frac{1}{2} = \frac{1}{2} - 2\varepsilon_n^2 \quad \text{or} \quad \varepsilon_{n+1} = -2\varepsilon_n^2.$$

When $n = 0$,

$$\varepsilon_1 = -2\varepsilon_0^2.$$

When $n = 1$,

$$\varepsilon_2 = -2\varepsilon_1^2 = -2(-2\varepsilon_0^2)^2 = -8\varepsilon_0^4.$$

When $n = 2$,

$$\varepsilon_3 = -2\varepsilon_2^2 = -2(-8\varepsilon_0^4)^2 = -2^7\varepsilon_0^8 = -\frac{1}{2}(2\varepsilon_0)^{2^3}.$$

Using the method of mathematical induction, we can show that

$$\varepsilon_n = -\frac{1}{2}(2\varepsilon_0)^{2^n}.$$

Indeed, assuming the above formula, we get:

$$\varepsilon_{n+1} = -2\left(-\frac{1}{2}(2\varepsilon_0)^{2^n}\right)^2 = \cdots = -\frac{1}{2}(2\varepsilon_0)^{2^{n+1}}.$$

Returning to the original variable u_n, we get

$$u_n = \frac{1}{2} + \varepsilon_n = \frac{1}{2} - \frac{1}{2}(2\varepsilon_0)^{2^n} = \frac{1}{2}[1 - (2u_0 - 1)^{2^n}],$$

which is an analytical solution of the logistic map when $r = 2$.

2.5 Periodic solutions

We have analyzed the existence and stability of fixed points in the logistic map. In particular, we found that when $r > 3$, there are no stable fixed points, and therefore orbits do not converge to any particular point. At the same time, all orbits remain bounded within the domain $[0, 1]$. It is possible that periodic solutions exist in the logistic map since they would satisfy both of the above conditions. Thus, our next task is to investigate whether periodic solutions are present in the logistic map.

Definition 2.6. A **two-periodic solution** is a solution that alternates between two values, that is, it satisfies $u_{n+2} = u_n$, or equivalently, $u_0 = u_2 = u_4 = \cdots$ and $u_1 = u_3 = \cdots$.

The corresponding conditions for a *three*-periodic solution is $u_{n+3} = u_n, \forall n$, and, more generally, for a k-periodic solution, $u_{n+k} = u_n, \forall n$.

Definition 2.7. Let $u_{n+1} = f(u_n)$. Then, $u_{n+2} = f(f(u_n)) = f^{(2)}(u_n)$ is a function of a function, which is called a **composite function** (or *two-fold composite*).

The two-fold composite function for the logistic map is a function giving u_{n+2} in terms of u_n

$$u_{n+2} = ru_{n+1}(1 - u_{n+1}) = r^2[u_n(1 - u_n)][1 - ru_n(1 - u_n)].$$

Thus,

$$f^{(2)}(u) = r^2[u(1 - u)][1 - ru(1 - u)].$$

Let us introduce a dynamical system generated by $f^{(2)}$ and described in terms of a new variable

$$x_{n+1} = f^{(2)}(x_n) = F(x_n),$$

that is,

$$x_{n+1} = r^2[x_n(1-x_n)][1-rx_n(1-x_n)]. \tag{2.8}$$

This is the two-composite logistic map. It is easy to see that if we set $x_0 = u_0$, then $x_1 = u_2$, $x_2 = u_4$, and, in general,

$$x_n = u_{2n}.$$

Alternatively, if we set $x_0 = u_1$, then $x_1 = u_3$, $x_2 = u_5$, and, in general,

$$x_n = u_{2n+1}.$$

Thus, two-periodic orbits in the logistic map correspond to fixed points in the map for x_n. In addition, fixed points of the logistic map are also fixed points in the map for x_n. Consequently, all fixed points of

$$x_{n+1} = F(x_n)$$

are either fixed points or two-periodic solutions of

$$u_{n+1} = f(u_n).$$

Let us try to find all these fixed points. They are solutions of the algebraic equation

$$x = r^2x(1-x)(1-rx+rx^2).$$

Since the fixed points of the logistic map are also fixed points of the composite map, $x = 0$ and $x = 1 - \frac{1}{r}$ are roots of this equation. After expanding all brackets and dividing by x, we can reduce the problem to finding the roots of the cubic polynomial

$$\begin{aligned}
1 &= (r^2 - r^2x)(1 - rx + rx^2) \\
&\Rightarrow \quad -r^3x^3 + r^3x^2 + r^3x^2 - r^2x - r^3x + r^2 - 1 = 0 \\
&\Rightarrow \quad r^3x^3 - 2r^3x^2 + (r^2 - r^3)x + (1 - r^2) = 0.
\end{aligned}$$

Using the fact that $x = 1 - 1/r$ is a root of this polynomial, we can factorize it (for example, by long division) as follows:

$$r\left(x - 1 + \frac{1}{r}\right)(r^2x^2 - (r^2 + r)x + r + 1) = 0.$$

Thus, the two additional fixed points, which correspond to two-periodic solutions of the logistic map, are the roots of the quadratic equation

$$r^2x^2 - (r^2 + r)x + r + 1 = 0.$$

As the two other fixed points are the same as those of the logistic map, $U_1 = 0$ and $U_2 = 1 - 1/r$, we denote these new roots as $U_{3,4}$ as follows:

$$U_{3,4} = \frac{r + 1 \pm \sqrt{(r+1)(r-3)}}{2r}, \tag{2.9}$$

which exist only when $r > 3$ (as otherwise they are complex). Thus, the occurrence of two-periodic solutions takes place at $r = 3$ when U_2 looses its stability. As U_3 and U_4 correspond to the two-periodic solution of the logistic map, $u_{n+1} = ru_n(1 - u_n)$, they should satisfy

$$\begin{aligned} U_4 &= rU_3(1 - U_3) \\ U_3 &= rU_4(1 - U_4). \end{aligned}$$

This can be directly verified as follows:

$$\begin{aligned} &rU_{3,4}(1 - U_{3,4}) \\ &= r\frac{r + 1 \pm \sqrt{(r+1)(r-3)}}{2r} \cdot \left(1 - \frac{r + 1 \pm \sqrt{(r+1)(r-3)}}{2r}\right) \\ &= \frac{r + 1 \pm \sqrt{(r+1)(r-3)}}{2} \cdot \frac{2r - (r + 1 \pm \sqrt{(r+1)(r-3)})}{2r} \\ &= \frac{r + (1 \pm \sqrt{(r+1)(r-3)})}{2} \cdot \frac{r - (1 \pm \sqrt{(r+1)(r-3)})}{2r} \\ &= \frac{r^2 - 1 \mp 2\sqrt{(r+1)(r-3)} - (r+1)(r-3)}{4r} \\ &= \frac{r + 1 \mp \sqrt{(r+1)(r-3)}}{2r} = U_{4,3}. \end{aligned}$$

Now, let us analyze the stability of the found two-periodic solution. Since U_3 is a fixed point of the map $u_{n+1} = F(u_n)$, it is stable if

$$|F'(u = U_3)| < 1.$$

A similar condition applies for stability of U_4, namely $|F'(u = U_4)| < 1$. Furthermore, since U_3 and U_4 correspond to the same two-periodic solution, they should be stable or unstable simultaneously.

Note that

$$F(u) = r^2 u(1 - u)(1 - ru + ru^2)$$

is a composite function, $F(u) = f^{(2)}(u)$. Therefore, differentiation of $F(u)$ can be performed using the chain rule

$$F'(u) = f'(f(u))f'(u),$$

or, at the particular point $u = U_3$,

$$\left.\frac{dF}{du}\right|_{u=U_3} = \left.\frac{df}{du}\right|_{u=f(U_3)=U_4} \left.\frac{df}{du}\right|_{u=U_3},$$

where $f(u) = ru(1-u)$ and $f'(u) = r - 2ru$.

Thus,

$$\begin{aligned} F'(u = U_3) &= f'(u = U_3)f'(u = U_4) = (r - 2rU_3)(r - 2rU_4) \\ &= r^2[1 - 2(U_3 + U_4) + 4U_3U_4]. \end{aligned}$$

It is easy to see that $F'(u = U_4) = F'(u = U_3)$. The two-periodic solution is stable when $-1 < F'(u = U_3) < 1$, or

$$-1 < r^2[1 - 2(U_3 + U_4) + 4U_3U_4] < 1.$$

Since $U_{3,4}$ are roots of

$$u^2 - \frac{r+1}{r}u + \frac{r+1}{r^2} = 0,$$

we can use the Vieta's formulas to find

$$U_3 + U_4 = \frac{r+1}{r}, \quad U_3U_4 = \frac{1+r}{r^2}.$$

Thus,

$$\begin{aligned} f'(U_3)f'(U_4) &= r^2(1 - 2(U_3 + U_4) + 4U_3U_4) \\ &= r^2\left(1 - 2\frac{r+1}{r} + 4\frac{r+1}{r^2}\right) = -(r^2 - 2r - 4) \\ &= -(r^2 - 2r + 1) + 5 = -(r-1)^2 + 5. \end{aligned} \tag{2.10}$$

Consequently, the two-periodic solution is stable if

$$-1 < -(r-1)^2 + 5 < 1.$$

After simple manipulations, we get

$$3 < r < 1 + \sqrt{6}.$$

Thus, the two-periodic solution is stable when r is in the range between 3 and $1 + \sqrt{6} \approx 3.45$.

Example 2.4. Find the value of the parameter r in the logistic map when its two-periodic solution is superstable.

Solution:
While the two-periodic solution is represented by the roots U_3 and U_4 of the equation $u = f^{(2)}(u)$, its stability is determined by the derivative

$$\left.\frac{df^{(2)}}{du}\right|_{u=U_3} = \left.\frac{df^{(2)}}{du}\right|_{u=U_4} = f'(U_3)f'(U_4) = -(r-1)^2 + 5,$$

as found in (2.10). Two-periodic solution is stable if this derivative is zero, i. e.,

$$-(r-1)^2 + 5 = 0.$$

This equation has two roots, $r = 1 \pm \sqrt{5}$, of which negative one can be rejected. Therefore, the two-periodic solution of the logistic map is superstable when

$$r = 1 + \sqrt{5} \approx 3.236.$$

2.6 Bifurcations

Looking at our results on the existence and stability of fixed points for the logistic map, we note that $U_1 = 0$ exists for all values of r and is stable when $r < 1$. At $r = 1$, U_1 loses its stability, and at the same time, the second fixed point, $U_2 = 1 - 1/r$, appears and is stable. Furthermore, at $r = 3$, the second fixed point, U_2, also becomes unstable.

Definition 2.8. Consider the map $u_{n+1} = f(u_n, C)$, where C is a parameter. Let's assume that U is a fixed point and is stable when $C < C_0$, and unstable when $C > C_0$. The value C_0 is called the **bifurcation value** of the parameter C.

Bifurcations are also associated with the appearance or disappearance of fixed points. For the logistic map, we have identified two bifurcations: At $r = 1$, the stability of U_1 changes and U_2 appears or disappears; and at $r = 3$, the stability of U_2 changes (see Figure 2.5). The bifurcation at $r = 3$ is called a *flip bifurcation* or *period-doubling bifurcation*, since it is the point where a two-periodic orbit appears (as will be shown).

Bifurcation theory is the study of the effect of parameters on the qualitative features of dynamical systems, particularly on the number and stability of their fixed points. The results of such studies are typically presented using a *bifurcation diagram*, which shows graphically, on the u–C plane, how the location and stability of fixed points U_i change with changes in the parameter C. A point (u, C), where a bifurcation takes place, is called a *bifurcation point*.

The bifurcation diagram for the logistic map is shown in Figure 2.5. Black lines indicate the locations of the fixed points of this map, depending on the value of the parameter r: Solid lines show the positions of stable fixed points, while dashed lines indicate unstable ones. The bifurcation diagram for the two-composite logistic map includes both

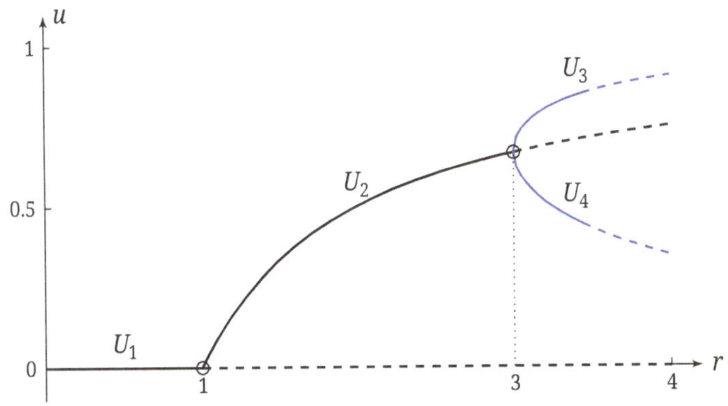

Figure 2.5: Bifurcation diagram for the logistic map. The fixed point $U_1 = 0$ exists for $r \in (0,4]$, is stable when $r < 1$, and becomes unstable when $r > 1$. The fixed point U_2 appears at $r = 1$, is stable for $1 < r < 3$, and becomes unstable when $r > 3$. The bifurcation points $(r,u) = (1,0)$ and $(r,u) = (3,2/3)$ are shown by small circles. The bifurcation diagram for the two-fold logistic map (2.8) includes additional branches shown in blue.

black and blue lines, with the blue lines corresponding to two additional fixed points of the composite map or to the two-periodic solution of the logistic map.

There are two bifurcation points on the bifurcation diagram shown in Figure 2.5. One occurs at the point $(r,u) = (1,0)$, where the fixed points U_1 and U_2 coincide and U_1 changes its stability. At this point, we observe a so-called *transcritical bifurcation*, which is defined as a bifurcation point, where fixed points collide and exchange stability.

At the bifurcation point $(r,u) = (3,2/3)$, the fixed point U_2 changes its stability without colliding with other fixed points (in the one-composite logistic map). Here, we observe a so-called *flip bifurcation*; this term refers to the formation of a periodic solution (flipping dynamics).

The bifurcation point $(r,u) = (3,2/3)$ has a different nature in the two-composite logistic map since we not only observe a change in the stability of U_2, but also the emergence of two new fixed points U_3 and U_4. For the two-composite logistic map, this point corresponds to a so-called *pitchfork bifurcation*.

A classification of bifurcations in one-variable systems will be given in the next chapter. Here, we simply note that there exists one further type of bifurcation, known as the *fold* (or *tangent*) bifurcation. It does not appear in the bifurcation diagram in Figure 2.5, but it often occurs in other one-dimensional maps.

2.7 Period-doubling cascade

As was shown in previous sections, the second fixed point U_2 in the logistic map exists when $r \geq 1$ and is stable if $r < 3$. When $r = 1$, the two fixed points coincide, $U_1 = U_2 = 0$, and the derivative $f'(u = U_2) = 1$. This derivative decreases with increasing r, so that

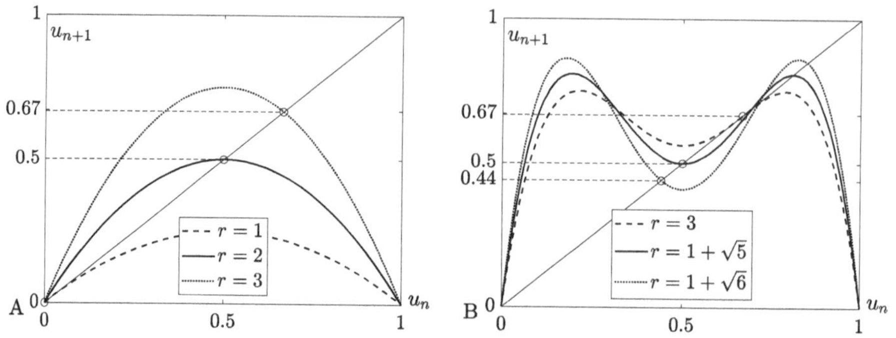

Figure 2.6: Panel A: plots of the logistic map with $r = 1$, 2, and 3. The location of the fixed point U_2, given by Eq. (2.7), is marked by a circle for each map. **Panel B:** plots of the two-composite logistic map with $r = 3$, $1 + \sqrt{5}$, and $1 + \sqrt{6}$. The location of the fixed point U_4, given by Eq. (2.9), is marked by a circle for each map.

when $r = 2$, we have $U_2 = 1/2$ and $f'(u = U_2) = 0$ (superstability). A further increase in the parameter r to $r = 3$ brings us to the point, where $U_2 = 2/3$ and $f'(u = U_2) = -1$ (flip bifurcation), indicating the loss of stability. These observations about U_2 are illustrated in Figure 2.6A.

When $r = 3$, a two-periodic solution emerges, which is represented by two fixed points, U_3 and U_4, of the two-composite logistic map (2.8). At $r = 3$, we have $U_3 = U_4 = U_2 = 2/3$ and $f'(u = U_3 = U_4) = 1$ (this is called a *pitchfork bifurcation*, which we will discuss later). Then, as we saw, at $r = 1 + \sqrt{5}$ we have $f'(u = U_4) = 0$, indicating superstability of the two-periodic solution. Further increase in r to $r = 1 + \sqrt{6}$ gives $f'(u = U_3) = f'(u = U_4) = -1$, indicating the loss of stability of the two-periodic solution (flip bifurcation). These observations about U_4 are illustrated in Figure 2.6B.

At $r = 1 + \sqrt{6}$, when the two-periodic solution becomes unstable, a stable four-periodic solution emerges, which exists for $r > 1 + \sqrt{6}$ and remains stable over a short range of r values. When the four-periodic solution becomes unstable, a eight-periodic solution appears, exhibiting the same behavior as the two-periodic and four-periodic solutions. This process repeats for all period-2^n solutions up to $n = \infty$, and is known as a *period-doubling cascade* (see Figure 2.7). The period-doubling cascade is associated with certain self-similar features, where the same properties are retained but at different scales.

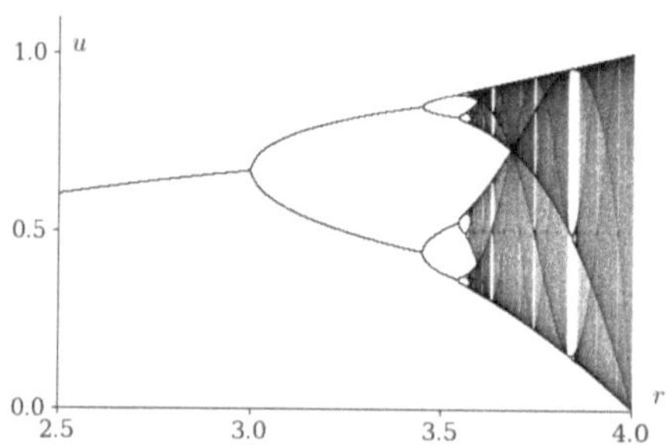

Figure 2.7: Orbit diagram showing all bifurcations, including the period-doubling cascades, in the logistic map. The first period-doubling cascade occurs for $1 < r \lesssim 3.57$. Solutions of all possible periodicities emerge in *Sharkovsky order* (2.11).

In the period-doubling cascade, self-similarity is characterized by two Feigenbaum constants, which define the scaling of this self-similarity. The distances between consecutive bifurcation points in the period-doubling cascade form a geometric progression with a common ratio given by the *first Feigenbaum constant*

$$\delta = \lim_{n\to\infty}\left(\frac{r_n - r_{n-1}}{r_{n+1} - r_n}\right) \approx 4.67,$$

where r_n is the bifurcation value associated with the emergence of a 2^n-periodic solution.

Similarly, the distances, Δu_n, between neighboring values of u in the 2^n-periodic solution—measured at the point when the solution becomes unstable (i. e., the vertical scaling in the bifurcation diagram)—decrease with increasing n, and also form a geometric progression. The common ratio of this progression is given by the *second Feigenbaum constant*

$$\alpha = \lim_{n\to\infty}\left(\frac{\Delta u_{n-1}}{\Delta u_n}\right) \approx 2.50.$$

The order in which solutions of different periodicities appear in the logistic map—illustrated by the bifurcation diagram in Figure 2.7 is governed by the so-called *Sharkovsky order* [7] as follows:

$$\begin{aligned} 1 < 2 < 4 < 8 < \cdots < 2^n < 2^{n+1} < \cdots \\ \cdots < 2^n \times (2k+1) < \cdots < 2^n \times 9 < 2^n \times 7 < 2^n \times 5 < 2^n \times 3 < \cdots \\ \cdots < 2^2 \times (2k+1) < \cdots < 2^2 \times 9 < 2^2 \times 7 < 2^2 \times 5 < 2^2 \times 3 < \cdots \\ \cdots < 2k+1 < \cdots < 9 < 7 < 5 < 3. \end{aligned} \tag{2.11}$$

Here, $m < l$ means that there exist values of the parameter r for which an m-periodic solution exists, while an l-periodic solution does not. In other words, if an m-periodic solution exists for a given value of r, then all periodic solutions whose periods appear to the left of m in the Sharkovsky order also exist (though they may be unstable), whereas periodic solutions whose periods appear to the right of m may not exist.

Infinite periodicity can be viewed as leading to unpredictability and marks the onset of *chaos*. The value of r corresponding to the emergence of a 2^∞-periodic solution can be approximated as the sum of a geometric progression. In this progression, each term represents the length of the stability interval for a 2^n-periodic solution, and each is linked to the previous by the first Feigenbaum constant δ. Specifically,

$$\begin{aligned} r_\infty &= r_0 + (r_1 - r_0) + (r_2 - r_1) + \cdots + (r_n - r_{n-1}) + \cdots \\ &\approx r_0 + (r_1 - r_0) + \frac{r_1 - r_0}{\delta} + \frac{r_1 - r_0}{\delta^2} + \cdots + \frac{r_1 - r_0}{\delta^n} + \cdots \\ &= r_0 + (r_1 - r_0)(1 + \delta^{-1} + \delta^{-2} + \cdots + \delta^{-n} + \cdots) \\ &= r_0 + (r_1 - r_0)\left(\frac{1}{1 - \delta^{-1}}\right). \end{aligned}$$

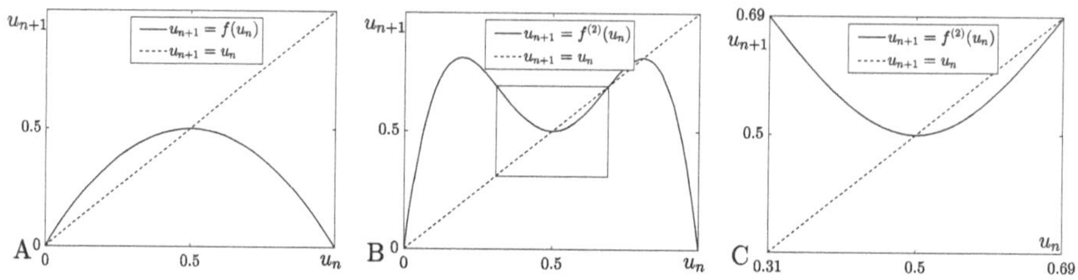

Figure 2.8: Towards the derivation of the second Feigenbaum constant. Panel A: the logistic map with $r = 2$, which has a superstable fixed point at $U_2 = 0.5$. **Panel B:** the two-composite logistic map with $r = 1 + \sqrt{5}$, which has a superstable fixed point at $U_4 = 0.5$. **Panel C:** an enlarged view of the region framed by thin lines in the center of Panel B. It is evident that Panel C contains a downscaled and rotated version of the plot in Panel A.

Taking $r_0 = 1$, $r_1 = 3$, and $\delta = 4.67$, we get

$$r_\infty \approx 1 + 2 \cdot \frac{4.67}{4.67 - 1} \approx 3.54.$$

The second Feigenbaum constant can be evaluated using the following considerations. By examining the plots of $f(u)$ and the two-composite map $f^{(2)}(u)$ shown in Figure 2.8, one can observe that the function $f(u)$ can be obtained from the central part of the composite map $f(f(u))$, after an appropriate rescaling and rotation. To express this observation mathematically, we introduce the variable $x = u - 1/2$. This variable represents the distance from the midpoint of the plots, which is a key point for both functions, since $f'(1/2) = 0$ and $f'(f(1/2)) = 0$. In terms of the variable x, we can express the scaling and rotation relation as

$$f(ax) = -af^{(2)}(-x), \tag{2.12}$$

where the two minus signs on the right-hand side account for rotation and the parameter a captures the scaling factor.

Let us consider the second-order approximation of $f(x)$ about $x = 0$

$$f(x) \approx f(0) + f'(0)x + \frac{1}{2!}f''(0)x^2 = f(0) + \frac{1}{2}f''(0)x^2$$

since $f'(0) = 0$. Using notations $b = f(0)$ and $-c = \frac{1}{2}f''(0)$, where both b and c are positive constants, we can rewrite this as

$$f(x) \approx b - cx^2.$$

We can use this for a second-order estimation of the relationship given by Eq. (2.12). For the function on the left-hand side of this equation, we obtain

$$f(ax) \approx f(0) + f'(0)\,ax + \frac{1}{2}f''(0)\,(ax)^2 = b - c(ax)^2.$$

For the function on the right-hand side, we calculate

$$f^{(2)}(-x) = f(f(-x)) \approx b - c(f(-x))^2 = b - c(b - cx^2)^2$$
$$= b - cb^2 + 2bc^2x^2 - c^3x^4.$$

Ignoring the fourth-order term c^3x^4 on the right-hand side, the equation reduces to

$$b - c(\alpha x)^2 = -\alpha(b - cb^2 + 2bc^2x^2).$$

By equating the constant terms, we obtain

$$b = \alpha b(cb - 1), \quad \text{which leads to } c = \frac{\alpha + 1}{\alpha b}.$$

Collecting the second-order terms, we obtain

$$-c\alpha^2 = -2\alpha bc^2, \quad \text{which leads to } c = \frac{\alpha}{2b}.$$

From these two expressions for c, we obtain

$$\alpha^2 - 2\alpha - 2 = 0.$$

The roots of this equation are

$$\alpha_{1,2} = 1 \pm \sqrt{1 + 2} = 1 \pm \sqrt{3},$$

giving

$$\alpha_1 = 1 + \sqrt{3} \approx 2.7321,$$

which is a meaningful solution and

$$\alpha_2 = 1 - \sqrt{3} \approx -0.732,$$

which should be rejected, as it is negative.

The actual value of the second Feigenbaum constant is approximately $\alpha \approx 2.5029$. Our first-order approximation $\alpha \approx 2.7321$ is not far off, indicating that the method captures the right qualitative behavior.

2.8 Quadratic maps

Definition 2.9. The map $u_{n+1} = f(u_n)$ is called a **quadratic map** if the map-generating function is a quadratic polynomial, i. e., $f(u) = au^2 + bu + c$.

Thus, the quadratic function is defined by the difference equation:

$$u_{n+1} = au_n^2 + bu_n + c,$$

which has three parameters a, b and c.

By setting $a = -r$, $b = r$, and $c = 0$, we recover the logistic map (2.6). Quadratic maps are *isomorphic*, that is, they can be transformed into one another via linear transformations. While this statement can be proved in general, we will illustrate the proof using the following example.

Example 2.5. Find a linear transformation of the form

$$u_n = \alpha y_n + \beta$$

to convert the logistic map (2.6) into the quadratic map, given by the equation

$$y_{n+1} = 1 - Ay_n^2. \tag{2.13}$$

Solution:

Let us substitute $u_n = \alpha y_n + \beta$ and $u_{n+1} = \alpha y_{n+1} + \beta$ into the logistic map:

$$y_{n+1} + \beta = r(\alpha y_n + \beta)(1 - \alpha y_n - \beta).$$

After expanding the right-hand side and isolating y_{n+1}, we obtain

$$y_{n+1} = -r\alpha y_n^2 + r(1 - 2\beta)y_n + \frac{\beta}{\alpha}(r - r\beta - 1).$$

We now want this to match Eq. (2.13), for which the leading coefficient is $-A$, the linear term is 0, and the free term is 1

$$\begin{aligned}
-r\alpha = -A \quad &\Rightarrow \quad \alpha = \frac{A}{r},\\
1 - 2\beta = 0 \quad &\Rightarrow \quad \beta = \frac{1}{2},\\
\frac{\beta}{\alpha}(r - r\beta - 1) = 1 \quad &\Rightarrow \quad \frac{r}{2A}\left(r - \frac{r}{2} - 1\right) = 1 \quad \Rightarrow \quad A = \frac{r}{2}\left(\frac{r}{2} - 1\right).
\end{aligned}$$

Thus, the linear transformation

$$u_n = \frac{A}{r}y_n + \frac{1}{2}$$

converts the logistic map (2.6) into the quadratic map (2.13).

This transformation also provides the relationship between the bifurcation values of parameters for these two maps, that is, $r = 1$ is the bifurcation value associated with the change in stability of $U_1 = 0$ and the emergence of $U_2 = 1 - \frac{1}{r}$ in the logistic map. This corresponds to the bifurcation occurring in the second map at

$$A = \frac{r}{2}\left(\frac{r}{2} - 1\right) = -\frac{1}{4}.$$

Two-periodic solutions arise in the logistic map when $r = 3$ and in the second map when $A(r = 3) = 3/4$.

2.9 Routes to chaos

As we can see from Figure 2.7, there are many values of the parameter r in the logistic equation for which chaos is observed. It appears that there are three ways in which we can enter the chaotic regime by changing the value of r in the logistic equation. We have seen that, when starting with low values of r and increasing it to around $r = 3.57$, we arrive at chaos via a *period-doubling cascade*. There are infinitely many period-doubling cascades, and further increases in the value of r will consequently take us through each of these cascades into the chaotic regime.

The second route to chaos is evident from an observation one can make by looking at Figure 2.7: As r increases, at certain values, chaotic solutions transform into stable 6-, 5-, or 3-periodic solutions. For example, the transition from chaos to a three-periodic solution takes place at a value of r near 3.83. There is a dynamic regime between periodic and chaotic solutions which is called an intermittent solution.

Definition 2.10. Intermittency is a sporadic switching between two qualitatively different behaviors, for example, a regime combining 3-periodic and chaotic solutions as illustrated in Figure 2.9B.

The explanation of the intermittent regime is as follows. At points where the plot of the map-generating function is very close to the diagonal line, the orbit can become trapped for awhile in the corridor between these two lines, resulting in nearly regular

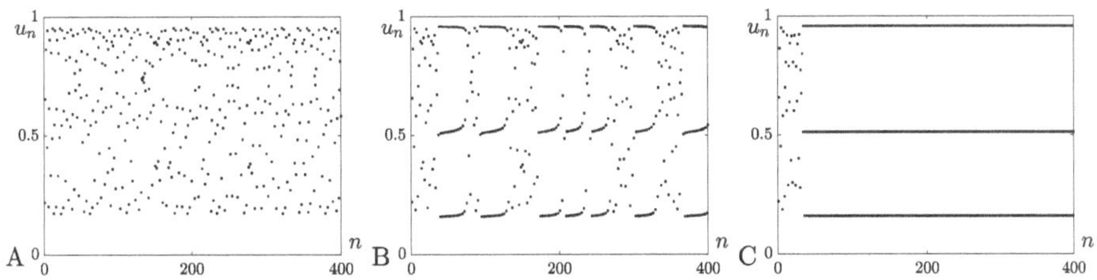

Figure 2.9: Illustration of chaotic, intermittent, and three-periodic regimes. Sequences of u_n generated by the logistic map (2.6) with initial condition $u_0 = 0.22$ and different values of the parameter r. **Panel A:** Chaotic regime, $r = 3.81$. **Panel B:** Intermittent regime, showing alternating chaotic and three-periodic bands, $r = 3.8277$. **Panel C:** Three-periodic regime, $r = 3.8285$.

inclusions within a chaotic solution. This is illustrated in Figure 2.10. At $r = 1 + \sqrt{8} \approx 3.8284$, the chaotic solution (for $r < 1 + \sqrt{8}$) transitions into a stable 3-periodic solution (for $r > 1 + \sqrt{8}$). Intermittency (orbits combining chaotic and 3-periodic behavior) is observed just before this transition, when the plot of the three-fold logistic map $f^{(3)}(u)$ comes very close to the diagonal line.

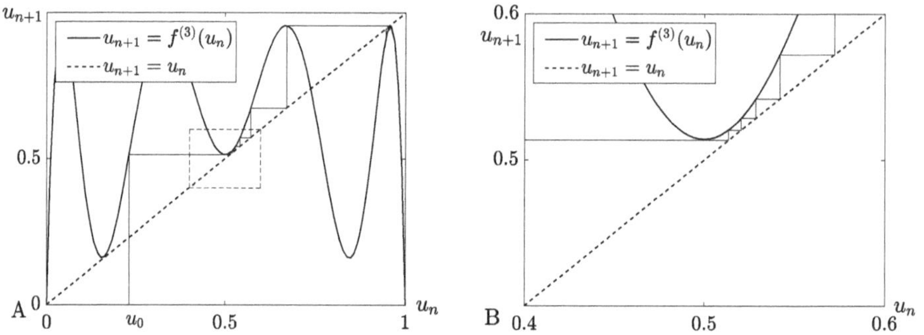

Figure 2.10: Intermittency in the logistic map. Panel A: The three-fold iterate of the logistic map, $f^{(3)}(u)$, is shown for $r = 3.826 < 1 + \sqrt{8}$, together with a cobwebbed trajectory (thin solid) starting from $u_0 = 0.228$. **Panel B:** An enlarged view of the region enclosed by the thin dashed borders in Panel A, illustrating how the trajectory becomes trapped between the central branch of $f^{(3)}(u)$ and the diagonal (dashed) line.

Let us estimate how long the orbit is trapped. Assume that the function $f(u)$ is at the shortest distance from the diagonal line when $u = U$. Then we can write $f(U) = U + \varepsilon$, where ε is a small parameter representing the vertical distance between the diagonal line and the function $f(u)$ at $u = U$. At the point where this distance is minimal, the curve has the same derivative as the diagonal line $f'(U) = 1$. The second-order approximation of $f(u)$ near the point U is

$$f(u) = U + \varepsilon + f'(U)(u - U) + \frac{1}{2!}f''(U)(u - U)^2 = u + \varepsilon + a(u - U)^2,$$

where $a = \frac{1}{2!}f''(U)$. Thus,

$$u_{n+1} = u_n + \varepsilon + a(u_n - U)^2.$$

Now, let us consider the change, Δu_n, in our variable u_n over one time step, from n to $n + 1$

$$\Delta u_n = u_{n+1} - u_n = \varepsilon + a(u_n - U)^2.$$

As Δu_n represents the change over a single time step, it can be interpreted as a rate of change and expressed as a derivative

$$\Delta u_n = \frac{du}{dn}; \quad \frac{du}{dn} = \varepsilon + a(u_n - U)^2.$$

The number of iterates required to pass through the corridor at $u = U$ can be estimated by the integral

$$\int_{u^-}^{u^+} \frac{du}{\varepsilon + a(u - U)^2} = \int_0^N dn = N,$$

where u^- and u^+ are values slightly below and above U, for example, $u^- = u_0 = U - \delta$ and $u^+ = u_N = U + \delta$. Thus,

$$N = \left[\frac{1}{\sqrt{a\varepsilon}} \arctan\left[(u - U)\sqrt{\frac{a}{\varepsilon}} \right] \right]_{u^-}^{u^+} \approx \frac{\pi}{\sqrt{a\varepsilon}},$$

with the estimate taken for $u_0 = -\infty$ and $u_N = \infty$. We see that the length of three-periodic inclusions in intermittent solution is proportional to $\varepsilon^{-1/2}$.

The third route to chaos is associated with the so-called *boundary crisis*, when unbounded orbits become bounded and chaotic. For the logistic map, if we allow $r > 4$, solutions become unbounded and $u_n \to -\infty$ as $n \to \infty$. If r is slightly above 4, there is a 'leak' around $u = 0.5$, where $f(u) > 1$, and if $u_{n+1} > 1$, it is easy to see that u_{n+2} becomes negative. When the value of the parameter r is gradually reduced and becomes $r \leq 4$, the orbits become bounded and chaotic (e. g.. when $r = 4$).

2.10 Nature of chaos

In one of the previous sections, we found an analytical solution for the logistic map with $r = 2$ when it has a superstable fixed point $U_2 = 0.5$. Here, we will find an analytical solution for the logistic map when $r = 4$, i. e.,

$$u_{n+1} = 4u_n(1 - u_n). \tag{2.14}$$

Using the substitution $u = \sin^2(\pi\theta)$, we replace u with θ. Note that since $|u| \leq 1$, the substitution is justified. Furthermore, we observe that $\sin^2(\pi\theta)$ is periodic with period 1

$$\sin^2(\pi(\theta + 1)) = \sin^2(\pi\theta),$$

and therefore we can consider the variable θ to be in the range $0 \leq \theta < 1$. In terms of the new variable, the logistic map transforms into

$$\begin{aligned} u_{n+1} &= 4u_n(1 - u_n) = 4\sin^2(\pi\theta_n)[1 - \sin^2(\pi\theta_n)] \\ &= 4\sin^2(\pi\theta_n)\cos^2(\pi\theta_n) = \sin^2(2\pi\theta_n) = \sin^2(\pi\theta_{n+1}). \end{aligned}$$

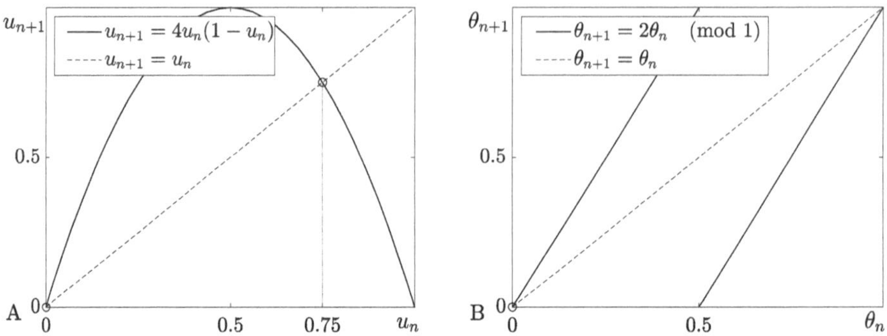

Figure 2.11: Plots of the logistic and doubling maps. Panel A: The logistic map (2.14), which has two fixed points, $U_1 = 0$ and $U_2 = 3/4$, both marked by circles. **Panel B:** The doubling map (2.16), which has a single fixed point, $\Theta_1 = 0$, marked by a circle.

Here, we have used the trigonometric identity $\sin(2x) = 2\sin(x)\cos(x)$. Thus, $u_{n+1} = \sin^2(\pi\theta_{n+1})$, where

$$\theta_{n+1} = 2\theta_n \pmod 1, \tag{2.15}$$

where (mod 1) is used to reflect the fact that θ is bounded between 0 and 1. Equation (2.15) represents the so-called *doubling map* (see Figure 2.11), which can also be expressed as

$$\theta_{n+1} = \begin{cases} 2\theta_n, & \text{if } 0 \le \theta_n < \frac{1}{2}, \\ 2\theta_n - 1, & \text{if } \frac{1}{2} \le \theta_n < 1. \end{cases} \tag{2.16}$$

Now, let θ_0 be defined as a fractional number represented as

$$\theta_0 = 0.a_1a_2a_3\ldots a_n\ldots$$

For what follows, it is convenient to use the binary representation of θ_0

$$\theta_0 = \frac{a_1}{2} + \frac{a_2}{2^2} + \frac{a_3}{2^3} + \cdots + \frac{a_n}{2^n} + \cdots,$$

where a_i, $\forall i$, is either 0 or 1. Now, θ_1 (or more generally, θ_n) is found by doubling θ_0 (or θ_{n-1}) and removing its integer part. Thus, each subsequent number in the orbit is obtained by removing the first digit of the previous number

$$\begin{aligned} \theta_1 &= 0.a_2a_3\ldots a_{n-1}\ldots \\ \theta_2 &= 0.a_3a_4\ldots a_{n-2}\ldots \\ &\vdots \\ \theta_k &= 0.a_{k+1}a_{k+2}a_{k+3}\ldots a_n\ldots \\ &\vdots \end{aligned}$$

If θ_0 is a rational number, then the solution either converges to 0 or is periodic, depending on whether the number is represented by a terminating or repeating sequence of digits. For example, if $\theta_0 = 0.(01)$, then the orbit is double-periodic

$$\theta_0 = 0.(01), \quad \theta_1 = 0.(10),$$
$$\theta_2 = \theta_0, \quad \theta_3 = \theta_1, \quad \text{and so on.}$$

Getting back to the decimal system, we obtain

$$\theta_0 = 0.(01) = \frac{1}{4} + \frac{1}{16} + \frac{1}{2^6} + \cdots = \frac{\frac{1}{4}}{1 - \frac{1}{4}} = \frac{1}{3},$$

and, similarly,

$$\theta_1 = 0.(10) = \frac{1}{2} + \frac{1}{8} + \frac{1}{2^5} + \cdots = \frac{\frac{1}{2}}{1 - \frac{1}{4}} = \frac{2}{3}.$$

There are two three-periodic solutions which, in the binary system, are represented by the periodic appearance of 110 or 001. The first one can be illustrated as follows:

$$\theta_0 = 0.(110), \quad \theta_1 = 0.(101), \quad \theta_2 = 0.(011),$$
$$\theta_3 = \theta_0, \quad \theta_4 = \theta_1, \quad \text{and so on,}$$

and the same solution applies in the decimal system

$$\theta_0 = \left(\frac{1}{2} + \frac{1}{4}\right) + \left(\frac{1}{2} + \frac{1}{4}\right)\frac{1}{2^3} + \left(\frac{1}{2} + \frac{1}{4}\right)\frac{1}{2^6} + \cdots = \frac{3}{4}\frac{1}{1 - 2^{-3}} = \frac{6}{7},$$
$$\theta_1 = \left(\frac{1}{2} + \frac{1}{8}\right) + \left(\frac{1}{2} + \frac{1}{8}\right)\frac{1}{2^3} + \left(\frac{1}{2} + \frac{1}{8}\right)\frac{1}{2^6} + \cdots = \frac{5}{8}\frac{1}{1 - 2^{-3}} = \frac{5}{7},$$
$$\theta_2 = \left(\frac{1}{4} + \frac{1}{8}\right) + \left(\frac{1}{4} + \frac{1}{8}\right)\frac{1}{2^3} + \left(\frac{1}{4} + \frac{1}{8}\right)\frac{1}{2^6} + \cdots = \frac{3}{8}\frac{1}{1 - 2^{-3}} = \frac{3}{7}.$$

If θ_0 is irrational, then the orbit is not periodic. Let, for example,

$$\theta_0 = \sqrt{2} \pmod 1 = 0.01101010000010011110 \ldots a_n \ldots$$

(in binary representation), for which we have no idea what a_n is when n is very large. Then,

$$\theta_n = 0.a_{n+1}a_{n+2} \ldots$$

is essentially an arbitrary number that cannot be predicted. This phenomenon underlies the chaotic dynamics observed in the doubling map, and consequently in the logistic map.

Summarizing the outcome of this section, we state that there are three kinds of orbits in the doubling map

$$\theta_{n+1} = 2\theta_n \pmod 1,$$

1. If θ_0 is a terminating decimal, then $\theta_n = 0$ for large enough n, and thus the orbit converges to the stationary (and unstable) fixed point $\theta = 0$.
2. If θ_0 is a repeating decimal, then the orbit is (eventually) periodic.
3. If θ_0 is irrational, then the orbit is chaotic.

It follows that the doubling map has a countable number of orbits ending at a fixed point or a periodic solution, while the majority of solutions are chaotic.

2.11 Structure of chaos

As we have seen, the orbits of the map $u_{n+1} = 4u_n(1 - u_n)$ are chaotic. Here, we check whether there is any structure in this chaos; namely, we will estimate how often different values of u occur in the chaotic solution. To estimate these frequencies, we will find the probability density function $p(u)$ that gives the probability that the points of the orbit lie in the interval between u and $u + du$

$$P(u < u_n < u + du) = p(u)du$$

Let us start with the probability density function for θ. Since all values of θ are equally likely, $p(\theta)$ is constant: $p(\theta) = c$. Since θ can only take values between 0 and 1, i. e., $0 \leq \theta \leq 1$,

$$\int_0^1 p(\theta)d\theta = 1 \Rightarrow \int_0^1 c d\theta = 1 \Rightarrow c = 1.$$

From Figure 2.12A, we see that the probability for u to lie between u and $u + du$ is equal to the probability for θ to lie between θ_1 and $\theta_1 + d\theta$, added to the probability for θ to lie between θ_2 and $\theta_2 - d\theta$. As the logistic function is symmetric with respect to $\theta = 0.5$ (i. e., $\theta_1 = 1 - \theta_2$), we can consider only $\theta \in [0, 0.5]$, and in this range we can state that

$$2p(\theta)d\theta = p(u)du.$$

Thus,

$$p(u) = \frac{2p(\theta)d\theta}{du} = \frac{2}{(du/d\theta)} = \frac{2}{2\pi \sin(\pi\theta)\cos(\pi\theta)}$$

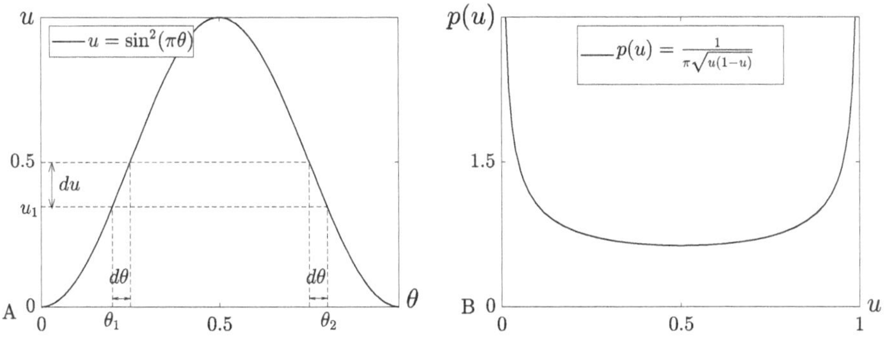

Figure 2.12: Panel A: The graph of $u = \sin^2(\pi\theta)$, used as the substitution that transforms the logistic map (2.14) into the doubling map (2.16). **Panel B:** The probability density function $p(u)$ describing the frequency of u-values in the chaotic regime of the logistic map (2.14).

since from $u = \sin^2(\pi\theta)$ we get $du/d\theta = 2\pi\sin(\pi\theta)\cos(\pi\theta)$. Also, from $u = \sin^2(\pi\theta)$, it follows that $\sin(\pi\theta) = \sqrt{u}$ and $\cos(\pi\theta) = \sqrt{1-u}$, and thus

$$p(u) = \frac{2}{2\pi\sin(\pi\theta)\cos(\pi\theta)} = \frac{1}{\pi\sqrt{u(1-u)}}.$$

The plot of this density function is shown in Figure 2.12B. We see that the probability density function $p(u)$ has a minimum at $u = 0.5$ and tends to infinity as u approaches zero or one, meaning that values of u near zero and one occur much more frequently than those around one half.

2.12 Second iterate of doubling map

The only fixed point of the doubling map $\theta_{n+1} = 2\theta_n \pmod 1$ is $\Theta = 0$, which is evident from its plot shown in Figure 2.11. The stability of the fixed point is determined by the derivative of the map-generating function

$$f(\theta) = 2\theta \pmod 1 \Rightarrow f'(\theta) = 2$$

everywhere except at $\theta = 0.5$, where the map has a discontinuity. Thus,

$$f'(\theta = 0) = 2 > 1,$$

and, therefore, the fixed point $\Theta = 0$ is unstable.

Consider the second iterate of the doubling map

$$\theta_{n+2} = 2\theta_{n+1} \pmod 1 = 2(2\theta_n) \pmod 1 = 4\theta_n \pmod 1.$$

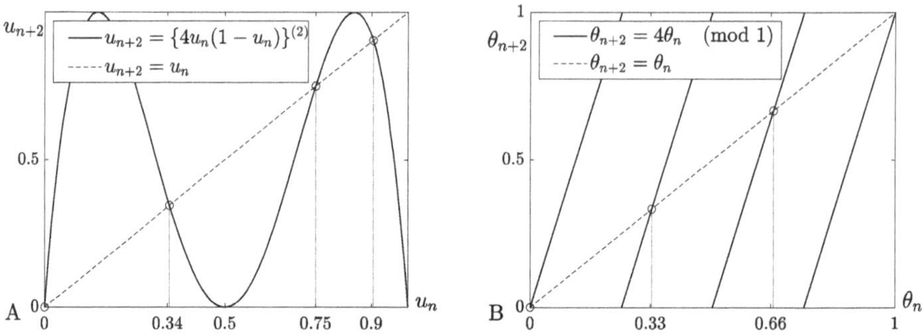

Figure 2.13: Plots of the two-composite logistic and doubling maps. Panel A: The two-fold composition of the logistic map (2.14), which has four fixed points, $U_1 = 0$, $U_2 = 3/4$, and $U_{3,4} = (5 \pm \sqrt{5})/8$, all marked by circles. **Panel B:** The two-fold composition of the doubling map (2.16), which has three fixed points, 0, 1/3, and 2/3, also marked by circles.

The plot of this map is shown in Figure 2.13B. The fixed points of this map are found as the roots of

$$4\theta - i = \theta$$

for $i = 0, 1, 2$, and 3. The fourth root (corresponding to $i = 3$) lies outside the half-open interval $[0, 1)$ and is therefore excluded. We thus have

$$\Theta_1 = 0, \quad \Theta_2 = \frac{1}{3}, \quad \Theta_3 = \frac{2}{3}.$$

While Θ_1 is a fixed point of the doubling map, Θ_2 and Θ_3 form its two-periodic solution.

Recall that we transformed the logistic map into the doubling map using the substitution: $u_n = \sin^2(\pi\theta_n)$. Also recall that the logistic map has two fixed points: $U_1 = 0$ and $U_2 = 1 - 1/r$. For $r = 4$, the second fixed point becomes $U_2 = 1 - \frac{1}{4} = \frac{3}{4}$ (see Figure 2.11A).

It is easy to see that two-periodic solution of the doubling map, given by $\Theta_2 = 1/3$ and $\Theta_3 = 2/3$, corresponds to the second fixed point of the logistic map

$$U_2 = \sin^2(\pi\Theta_2) = \sin^2\left(\frac{\pi}{3}\right) = \frac{3}{4}.$$

And also

$$U_2 = \sin^2(\pi\Theta_3) = \sin^2\left(\frac{2\pi}{3}\right) = \frac{3}{4}.$$

It appears that the two-periodic solutions of the logistic map correspond to four-periodic solutions of the doubling map, and, more generally, that 2^n-periodic solutions of the logistic map correspond to 2^{n+1}-periodic solutions of the doubling map.

2.13 Tent map

Earlier in this chapter, in Example 2.3, we considered a particular version of the tent map. Here, we will examine a more general form of the tent map, which involves the use of a parameter.

Definition 2.11. The discrete-time dynamical system given by the following difference equation:

$$x_{n+1} = \begin{cases} 2\mu x_n, & 0 \le x_n < \frac{1}{2}, \\ 2\mu(1 - x_n), & \frac{1}{2} \le x_n \le 1, \end{cases}$$

where $0 < \mu \le 1$, is called a **tent map**.

The tent map is a piecewise-linear version of the logistic map. The dynamics of this map depend on the parameter μ in much the same way as, in the logistic map, they depend on the parameter r.

Fixed points are roots of the equation $x_{n+1} = f(x_n) = x$. If $x < 1/2$, then $x = f(x)$ implies $x = 2\mu x$, which has the only solution $X_1 = 0$ if $\mu \neq 1/2$, or infinitely many solutions if $\mu = 1/2$.

If $x > 1/2$, then $x = f(x)$ implies $x = 2\mu(1-x)$, which has the solution $X_2 = 2\mu/(1+2\mu)$, satisfying the condition $x > 1/2$ only when $\mu > 1/2$.

Thus, when $\mu < 1/2$ there is only one fixed point $X_1 = 0$, and when $\mu > 1/2$ there are two fixed points: $X_1 = 0$ and $X_2 = 2\mu/(1 + 2\mu)$.

Stability of fixed points. For $x < 1/2$, $df/dx = 2\mu$, and thus $X_1 = 0$ is stable if $\mu < 1/2$ and unstable when $\mu > 1/2$. For $x > 1/2$, $df/dx = -2\mu$, and since $X_2 = 2\mu/(1 + 2\mu)$ exists only when $\mu > 1/2$, it is unstable.

Two-periodic solutions are found as an extra pair of fixed points for the two-composite map, with the map-generating function given by $f^{(2)}(x)$. For the analysis of the function $f^{(2)}(x)$, we note the symmetry $f(x) = f(1 - x)$, which implies the symmetry $f^{(2)}(x) = f^{(2)}(1 - x)$ (this symmetry about $x_0 = 1/2$ is visible in both panels of Figure 2.14). Therefore, we can restrict our analysis to the case $x_0 < 1/2$ and then extend it to the branch with $x_0 > 1/2$.

If $\mu < 1/2$, then $\forall x_0 \in [0, 1/2]$ we have $x_1 = \mu x_0 < 1/2$, and hence $x_2 = 2\mu x_1 = 4\mu^2 x_0 (< 1/2)$. Thus, $f^{(2)}(x) = 4\mu^2 x$ for the left branch of the function, which, taking into account the symmetry noted above, can be extended to the right branch (see Figure 2.14B).

If $\mu > 1/2$, then $x_1 = 2\mu x_0 < 1/2$ if $x_0 < 1/(4\mu)$, and $x_1 > 1/2$ if $1/(4\mu) < x_0 \le 1/2$. Therefore, the left branch of the function $f(x)$ splits into two branches (given by linear functions), with a boundary at $x = 1/(4\mu)$, where $x_1 = 2\mu/(4\mu) = 1/2$ and $x_2 = \mu$ at this boundary.

To complete the analysis of $f^{(2)}(x)$ for $x_0 \in [0, 1/2]$, we only need to find x_2 at $x_0 = 1/2$. For $x_0 = 1/2$, we have $x_1 = \mu$, and consequently, $x_2 = 2\mu(1 - \mu)$. Using this result,

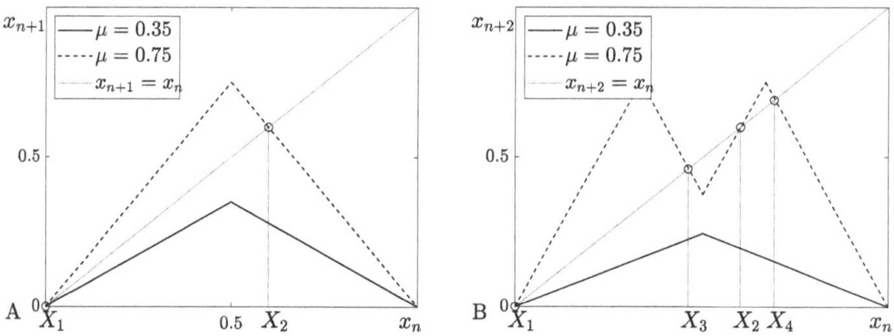

Figure 2.14: Tent map and its two-fold composition. Panel A: The tent map for $\mu < 1/2$ (solid line) and for $\mu > 1/2$ (dashed line). **Panel B:** The two-fold composition $f \circ f$ of the tent map, shown for $\mu < 1/2$ (solid line) and $\mu > 1/2$ (dashed line). Fixed points are marked by circles.

and extending it to the interval $1/2 < x \leq 1$, we have produced the sketch of $f^{(2)}(x)$ in Figure 2.14.

Analytical representation of $f^{(2)}(x)$ in the case of $0 < \mu \leq 0.5$ is the following:

$$f^{(2)}(x) = \begin{cases} 4\mu^2 x, & 0 \leq x < \frac{1}{2}, \\ 4\mu^2(1-x), & \frac{1}{2} \leq x \leq 1. \end{cases}$$

And, if $1/2 < \mu \leq 1$, then $f^{(2)}(x)$ is a piecewise linear function composed of the four branches

$$f^{(2)}(x) = \begin{cases} 4\mu^2 x, & 0 \leq x < \frac{1}{4\mu}, \\ 2\mu(1-2\mu x), & \frac{1}{4\mu} \leq x < \frac{1}{2}, \\ 2\mu(1-2\mu(1-x)), & \frac{1}{2} \leq x < \frac{4\mu-1}{4\mu}, \\ 4\mu^2(1-x), & \frac{4\mu-1}{4\mu} \leq x \leq 1. \end{cases}$$

Getting back to the fixed points of $f^{(2)}(x)$, we note that for $\mu < 1/2$, it does not have a fixed point other than $X_1 = 0$. Hence, there is no two-periodic solution in the original dynamical system generated by $f(x)$.

For $\mu > 1/2$, in addition to $X_1 = 0$ and $X_2 = \frac{2\mu}{1+2\mu}$, there are two extra fixed points on the second and fourth branches, where $\frac{1}{4\mu} \leq x \leq \frac{1}{2}$ and $\frac{4\mu-1}{4\mu} \leq x \leq 1$, given by $X_3 = \frac{2\mu}{1+4\mu^2}$ and $X_4 = \frac{4\mu^2}{1+4\mu^2}$, forming a two-periodic solution. The derivative of $f^{(2)}(x)$ is defined as

$$\frac{df^{(2)}(x)}{dx_0} = \frac{dx_2}{dx_1}\frac{dx_1}{dx_0} = (\pm 2\mu)(\pm 2\mu),$$

hence

$$\left|\frac{df^{(2)}}{dx}\right| = 4\mu^2 > 1$$

since $\mu > 1/2$.

Altogether, for $\mu > 1/2$, the function $f^{(2)}(x)$ has no stable fixed points, and hence the tent map has no stable two-periodic solutions.

2.14 Lyapunov exponent

As we have seen earlier, chaotic behaviour in the doubling map (2.15) arises when the initial value of the dynamic variable, θ_0, is irrational. In contrast, if θ_0 is rational, the system eventually converges to the fixed point $U_1 = 0$ or enters a periodic orbit. This observation extends to the logistic map (2.14), where chaos emerges when $\arcsin(\sqrt{u_0})/\pi$ is irrational. In general, the formation of chaos implies that orbits neither converge to a fixed point nor settle into a periodic solution, and they do not diverge to infinity. Convergence to a fixed point or periodic orbit means that trajectories eventually approach one another, while divergence to infinity implies unbounded behavior. Therefore, a necessary condition for chaos is the divergence of nearby orbits while remaining bounded—this property is commonly referred to as *sensitivity to initial conditions*. This sensitivity is quantified by the *Lyapunov exponent*, which we introduce in this section.

Consider two orbits, one starting at the point x_0 and the other at y_0, and assume that these two points are very close: $y_0 = x_0 + \varepsilon_0$, where $|\varepsilon_0| \ll 1$. We want to determine whether these two orbits converge or diverge. The distance between the two orbits after n iterates is ε_n as follows:

$$|\varepsilon_n| = |y_n - x_n| = \left|f^{(n)}(y_0) - f^{(n)}(x_0)\right| = \left|f^{(n)}(x_0 + \varepsilon_0) - f^{(n)}(x_0)\right|.$$

Using the first order approximation,

$$f^{(n)}(x_0 + \varepsilon_0) = f^{(n)}(x_0) + \left.\frac{df^{(n)}}{dx}\right|_{x=x_0} \varepsilon_0,$$

we obtain

$$|\varepsilon_n| = \left|\frac{df^{(n)}}{dx}\right|_{x=x_0} |\varepsilon_0|.$$

To each derivative $\frac{df^{(n)}}{dx}|_{x=x_0}$, we can associate a number λ_n such that

$$\left|\frac{df^{(n)}}{dx}\right|_{x=x_0} = \exp(\lambda_n n).$$

The limit

$$\lambda = \lim_{n\to\infty} \lambda_n = \lim_{n\to\infty} \frac{1}{n} \ln \left| \frac{df^{(n)}}{dx} \right|_{x=x_0}$$

(if it exists) is called the **Lyapunov exponent**. There is good reason to expect that this limit exists. If $\frac{df^{(n)}}{dx}|_{x=x_0}$ grows (or decays) exponentially with n, then λ is finite. If the derivative changes (grows or decays) faster than exponentially, then $|\lambda| = \infty$, and if slower, then $\lambda = 0$. We also expect that λ is independent of x_0 for almost all x since the expression $\frac{df^{(n)}}{dx}|_{x=x_0}$ is a product of df/dx evaluated at various iterates of x, effectively averaging or "smoothing" local variations over the entire domain.

Obviously, $\lambda > 0$ implies that $\varepsilon_n \to \infty$ when $n \to \infty$, meaning that nearby orbits diverge over time. This corresponds to sensitivity to initial conditions, a hallmark of chaos. Conversely, if $\lambda < 0$, then $\varepsilon_n \to 0$ indicating that nearby orbits converge, and there is no sensitivity to initial conditions. Since $f^{(n)}(x) = f(f^{n-1}(x))$, we can apply the chain rule iteratively to obtain

$$\left.\frac{df^{(n)}}{dx}\right|_{x=x_0} = \left.\frac{df}{dx}\right|_{x=f^{n-1}(x_0)} \left.\frac{df^{(n-1)}}{dx}\right|_{x=x_0}.$$

Applying this recursively yields

$$\left.\frac{df^{(n)}}{dx}\right|_{x=x_0} = \left.\frac{df}{dx}\right|_{x=f^{(n-1)}(x_0)} \left.\frac{df}{dx}\right|_{x=f^{(n-2)}(x_0)} \cdots \left.\frac{df}{dx}\right|_{x=f(x_0)} \left.\frac{df}{dx}\right|_{x=x_0}.$$

This is often expressed using more compact notation

$$\frac{df^{(n)}}{dx_0} = f'(x_{n-1})f'(x_{n-2})\ldots f'(x_0),$$

or even

$$\frac{dx_n}{dx_0} = \frac{dx_n}{dx_{n-1}} \frac{dx_{n-1}}{dx_{n-2}} \cdots \frac{dx_1}{dx_0}.$$

Returning to the definition of the Lyapunov exponent, we can write

$$\ln\left|\frac{df^{(n)}}{dx_0}\right| = \ln\left|f'(x_{n-1})f'(x_{n-2})\ldots f'(x_0)\right| = \ln\left|f'(x_{n-1})\right| + \cdots + \ln\left|f'(x_0)\right|.$$

Therefore, the Lyapunov exponent becomes

$$\lambda = \lim_{n\to\infty} \left(\frac{1}{n} \sum_{i=0}^{n-1} \ln\left|f'(x_i)\right| \right).$$

This formula allows us to calculate the Lyapunov exponent for a wide range of maps, enabling conclusions about the sensitivity of orbits to initial conditions and the potential for chaotic behaviour in these systems.

Example 2.6. Assume that $f'(x)$ is constant, i. e., it takes the same value for all x as is the case in the doubling map. Then,

$$\begin{aligned}\lambda &= \lim_{n\to\infty}\frac{1}{n}\underbrace{(\ln|f'(x)|+\ln|f'(x)|+\cdots+\ln|f'(x)|)}_{n\text{ terms}}\\ &= \lim_{n\to\infty}\left(\frac{1}{n}n\ln|f'(x)|\right)=\ln|f'(x)|.\end{aligned}$$

If $|f'(x)| < 1$, the Lyapunov exponent is negative, i. e.,

$$\lambda = \ln|f'(x)| < 0,$$

which implies insensitivity to initial conditions—nearby orbits converge. Therefore, the system does not exhibit chaos.

Conversely, if $|f'(x)| > 1$, then

$$\lambda = \ln|f'(x)| > 0,$$

which implies sensitivity to initial conditions—nearby orbits diverge. Provided the orbits remain bounded, this is a hallmark of chaotic behavior.

Example 2.7. Find the Lyapunov exponent for the logistic map

$$u_{n+1} = ru_n(1-u_n).$$

Solution:

Following the analysis of the logistic map made in Section 2.4, we consider its two fixed points. For the fixed point $U_1 = 0$, we have

$$f'(U_1 = 0) = r.$$

If $0 < r < 1$, then $U_1 = 0$ is stable, and we know that all orbits in this case converge to this fixed point, and thus

$$\lambda = \lim_{n\to\infty}\left(\frac{1}{n}\sum_{i=0}^{n-1}\ln|f'(u_i)|\right) = \ln|f'(U_1=0)| = \ln r < 0.$$

If $1 < r < 3$, then $U_2 = 1 - 1/r$ is stable (since $f'(U_2 = 1-1/r) = 2-r$ is between -1 and 1), and all orbits in this case converge to this fixed point. Thus,

$$\lambda = \lim_{n\to\infty}\left(\frac{1}{n}\sum_{i=0}^{n-1}\ln|f'(u_i)|\right) = \ln|f'(u = 1-1/r)| = \ln|2-r| < 0.$$

If $3 < r < 1+\sqrt{6}$, the two-periodic solution (U_3, U_4) is stable, that is, $|f'(U_3)f'(U_4)| < 1$. For the Lyapunov exponent, we have

$$\begin{aligned}\lambda &= \lim_{n\to\infty}\left(\frac{1}{n}\sum_{i=0}^{n-1}\ln|f'(U_i)|\right)\\ &= \lim_{k\to\infty}\left(\frac{1}{2k}\sum_{i=0}^{k}(\ln|f'(U_{2i})| + \ln|f'(U_{2i+1})|)\right)\\ &= \frac{1}{2}(\ln|f'(U_3)| + \ln|f'(U_4)|) = \frac{1}{2}\ln|f'(U_3)f'(U_4)| < 0.\end{aligned}$$

Generally, the Lyapunov exponent depends on the orbit (i. e., on the value of u_0); however, for relatively simple maps (including the logistic map), this is not the case. In Figure 2.15, it is shown how the Lyapunov exponent changes with the parameter r in the logistic map. It is evident that the Lyapunov exponent is negative if, for the given value of r, there is a stable periodic orbit, and positive when there are no stable solutions. The Lyapunov exponent reaches its maximum when $r = 4$.

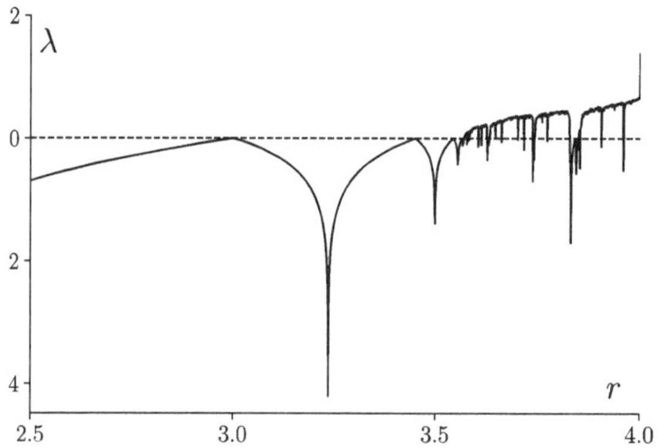

Figure 2.15: Lyapunov exponent as a function of the parameter r for the logistic map (2.6). Positive values of the exponent correspond to chaotic regimes.

If $r = 4$, then, following the substitution suggested in Section 2.10 for transforming the logistic map into the doubling map, we have

$$\left|\frac{df^{(n)}}{du_0}\right| = \left|\frac{df^{(n)}}{d\theta_0} \times \frac{d\theta_0}{du_0}\right| = 2^n\left|\frac{d\theta_0}{du_0}\right|$$

almost everywhere, and

$$\lambda = \lim_{n\to 0}\frac{1}{n}\ln\left|2^n\frac{d\theta_0}{du_0}\right| = \ln 2 > 0,$$

and, since the orbits are bounded, we have *chaos*.

2.15 Summary

In this chapter, we have focused on one-variable discrete-time models of dynamical systems represented by difference equations or maps. We have omitted a detailed consideration of the properties of these equations, which can be found in a number of textbooks (see, for example, [9]). Instead, we have concentrated on the qualitative analysis of one-variable autonomous maps since this is most relevant to the study of dynamical systems, particularly in the modeling of biological processes [5, 17].

We have identified several types of solutions to autonomous maps, $u_{n+1} = f(u_n)$, which include:

- stationary solutions, represented by fixed points, which are given by the roots of the algebraic equation $f(x) = x$.
- solutions converging to or diverging from fixed points. The existence of such solutions is closely associated with the stability of fixed points, as defined by the absolute value of the derivative $f'(x)$ at the fixed point, namely, by the condition $|f'(x)| < 1$.
- k-periodic solutions, defined by the condition $u_{n+k} = u_n$, and represented by the set of k values of u as follows: $U_1, U_2 = f(U_1), \dots, U_k = f(U_{k-1}) = U_1$.
- solutions converging to or diverging from periodic solutions. The existence of the latter is associated with the stability of the periodic solution, as defined by the derivatives $f'(U_i)$ along the periodic orbit, namely, by the condition $|f'(U_1) \cdot f'(U_2) \cdot \dots \cdot f'(U_k)| < 1$.
- chaotic solutions, which exist when solutions are bounded, and there are no stable fixed or periodic solutions. The condition for the existence of chaotic solutions can be formulated in terms of the Lyapunov exponent, namely, that the Lyapunov exponent along the chaotic solution is positive.

The calculation of the Lyapunov exponent can, in some cases, be a difficult and time-consuming task. Fortunately, for certain classes of map-generating functions $f(u)$, it can be concluded immediately that no chaotic solutions can arise.

Definition 2.12. In the dynamical system given by the map $u_{n+1} = f(u_n)$, there is no chaos if $f(u)$ is invertible. This statement is referred to as **Arnold's conjecture**.

The motivation behind this conjecture is that sensitivity to initial conditions is associated with a loss of information about the initial state, which, in turn, is ensured by non-invertibility. Note that Arnold's conjecture is not applicable to multidimensional maps. In Chapter 5 we will illustrate this by means of an invertible two-dimensional map that nevertheless exhibits chaotic dynamics.

Example 2.8. Analyze how the dynamics of the map

$$u_{n+1} = au_n + b \pmod 1$$

depend on the parameters a and b.

Solution:
Note that adding any integer to b leaves the map unaltered, and therefore we can consider only $0 \le b < 1$. Furthermore, we can see that if $a \ne 1$, then all fixed points are given by

$$U_i = \frac{b+i}{1-a}$$

for all integer values of i such that

$$0 < \frac{b+i}{1-a} < 1.$$

Also, we note that if $a = 1$, there are no fixed points if $b \ne 0$, and infinitely many fixed points otherwise. As the derivative of the map-generating function is equal to a (wherever it is defined), all fixed points, as well as all periodic solutions, are unstable if $|a| > 1$ and stable if $|a| < 1$. In the case of $a = 1$, all solutions are periodic if b is a rational number and called *quasiperiodic* if b is irrational.

The material covered in this chapter can also be found in other textbooks aimed at physics [19] or mathematics [7, 15] students. Further problems for practicing the mathematical techniques presented here can be found in [8].

2.16 Exercises

Problem 2.1. Consider the dynamical system defined by the map $x_n = f(x_{n-1})$ in the domain $[0, 1]$. For each of the following cases:
(a) $f(x) = x^3$;
(b) $f(x) = 4x(1-x)$,

answer the following:
(i) Sketch the function $f(x)$ and discuss whether $f(x)$ is invertible;
(ii) Use a calculator to compute x_n for $n = 1, 2, 3$ with the initial condition $x_0 = 0.7071$;
(iii) Sketch the composition $f(f(x))$;
(iv) Find the fixed points and analyze their stability.

Problem 2.2. Consider the tent map, or the map $x_{n+1} = f(x_n)$ given by the function $f(x)$ defined on $x \in [0, 1]$:

$$f(x) = \begin{cases} 2x, & 0 \le x < \frac{1}{2}, \\ 2(1-x), & \frac{1}{2} \le x \le 1. \end{cases}$$

(i) Find the fixed points of this system and show that they are unstable.

(ii) Find $F(x) = f(f(x))$. Sketch $F(x)$, find its fixed points and analyze their stability. Hence, for the map $x_{n+1} = f(x_n)$, describe the two-periodic solution and its stability.

Problem 2.3. For the map

$$x_{n+1} = x_n^3 - \mu x_n, \tag{2.17}$$

defined for $x_n, \mu \in \mathbb{R}$, find the fixed points and determine their stability. Identify the bifurcation values of μ and the types of bifurcation occurring at these values.

Problem 2.4. For the map $x_{n+1} = 1 - ax_n^2$ defined for $x_n, a \in \mathbb{R}$, $a \neq 0$, find the conditions for the existence and stability of

(i) All fixed points;
(ii) Two-periodic solutions.

In the course of your work, indicate and classify all bifurcations associated with the existence and stability of fixed points and two-periodic solutions.

Problem 2.5. Given the following numerical estimates of parameter r for superstable points of the logistic map (2.6), show that the period-doubling sequence is approximately self-similar and estimate the Feigenbaum constant δ for:

$$\begin{aligned} n &= 4: & r &= 3.49856170 \\ n &= 8: & r &= 3.55464086 \\ n &= 16: & r &= 3.56666738 \\ n &= 32: & r &= 3.56924353 \\ n &= 64: & r &= 3.56979529. \end{aligned}$$

Also, taking into account that these r-values form approximately a geometric series, estimate the r-value of the accumulation point of the period-doubling sequence, where chaos begins, namely, show that the value of r corresponding to $n = \infty$ is approximately 3.57.

Problem 2.6. For the dynamical system described by the map

$$x_{n+1} = \begin{cases} x_n + \frac{1}{2}, & 0 \le x_n \le \frac{1}{2}, \\ 2 - 2x_n, & \frac{1}{2} < x_n \le 1, \end{cases}$$

estimate the Lyapunov exponent and deduce whether the behaviour of the system is chaotic or regular.

Problem 2.7. For the dynamical systems defined by the maps

(a) $x_{n+1} = x_n + \pi/4 \pmod 1$;

(b) $x_{n+1} = \sqrt{3}x_n \pmod 1$;
(c) $x_{n+1} = x_n^{1/4} \pmod 1$,

find the Lyapunov exponent and deduce whether the behavior of the system is chaotic or regular.

2.17 Solutions to exercises

Solution to Problem 2.1.

(a) $f(x) = x^3$
  (i) Graphs of $f(x)$ and $f^{(2)}(x)$ are shown in Figure 2.16A.
  (ii) The calculator gives us $x_1 = 0.3535$, $x_2 = 0.0442$, and $x_3 = 0.00009$.
  (iii) $f(x) = x^3$ is invertible since it is a strictly increasing function (there is a one-to-one correspondence).
  (iv) Its fixed points are $X_1 = 0$ and $X_2 = 1$. Since $f'(x) = 3x^2$, it follows that X_1 is superstable ($f'(X_1) = 0$), and X_2 is unstable ($f'(X_2) = 3 > 1$).

(b) $f(x) = 4x(1-x)$.
  (i) Graphs of $f(x)$ and $f^{(2)}(x)$ are shown in Figure 2.16B.
  (ii) The calculator provides us with $x_1 = 0.8284$, $x_2 = 0.5685$, and $x_3 = 0.9812$.
  (iii) $f(x) = 4x(1-x)$ is not invertible since it is not monotonic.
  (iv) Its fixed points are $X_1 = 0$ and X_2 such that

$$4X_2(1 - X_2) = X_2 \Rightarrow 4 - 4X_2 = 1,$$

hence, $X_2 = 3/4$. Since $f'(x) = 4 - 8x$, it follows that both X_1 and X_2 are unstable ($f'(X_1) = 4 > 1$ and $f'(X_2) = -2 < -1$).

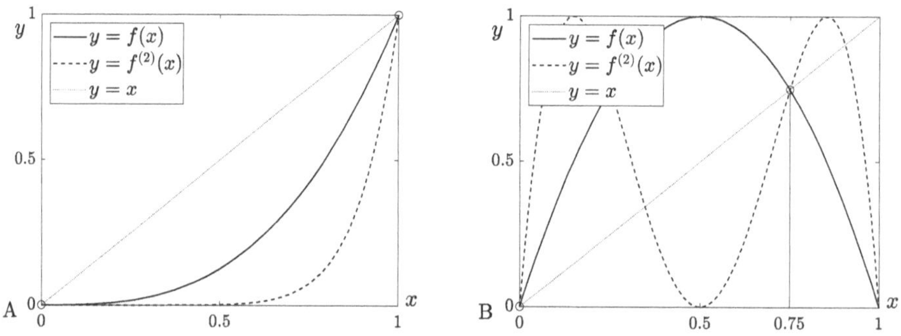

Figure 2.16: Plots of maps and their two-fold compositions. Panel A: $f(x) = x^3$ (dashed) and its two-fold composition $f(f(x)) = x^9$ (dotted). **Panel B:** $f(x) = 4x(1-x)$ (dashed) and its two-fold composition $f(f(x)) = 16x(1-x)(1-4x(1-x))$ (dotted).

Solution to Problem 2.2.

(i) Fixed points are the points of intersection between the map-generating function $y = f(x)$ and the diagonal line $y = x$.
For the first branch of the tent map, $x = 2x$, which gives the fixed point $X_1 = 0$. For the second branch of the tent map, $x = 2(1-x)$, which gives the fixed point $X_2 = 2/3$. Since $f'(x = 0) = 2 > 1$ and $f'(x = 2/3) = -2 < -1$, both fixed points are unstable.

(ii) The second iterate of the tent map is given by the map-generating function having four branches as follows:

$$F(x) = f(f(x)) = \begin{cases} 4x, & 0 \le x < \frac{1}{4}, \\ 2-4x, & \frac{1}{4} \le x < \frac{1}{2}, \\ 4x-2, & \frac{1}{2} \le x < \frac{3}{4}, \\ 4-4x, & \frac{3}{4} \le x \le 1. \end{cases}$$

Fixed points as follows:
- On the first branch: $x = 4x$ or $X_1 = 0$.
- On the second branch: $x = 2 - 4x$ or $X_2 = \frac{2}{5}$.
- On the third branch: $x = 4x - 2$ or $X_3 = \frac{2}{3}$.
- On the fourth branch: $x = 4 - 4x$ or $X_4 = \frac{4}{5}$.

All fixed points are unstable since $|F'(x = x_i)| = 4 > 1$.
Two-periodic solution is composed of $X_2 = 2/5$ and $X_4 = 4/5$, and it is unstable.

Solution to Problem 2.3. Fixed points are roots of the equation $x = x^3 - \mu x$. One root is $X_1 = 0$, and two other roots are found from $1 = x^2 - \mu$, or

$$X_{2,3} = \pm\sqrt{\mu + 1},$$

which exist when $\mu > -1$.

To analyze stability, we need the derivative: $f'(x) = 3x^2 - \mu$.

For the fixed point $X_1 = 0$ we have $f'(x = X_1) = -\mu$, and thus X_1 is stable if $|\mu| < 1$ and unstable otherwise. At $\mu = 1$ we have a flip bifurcation, and at $\mu = -1$ we may have a tangent, transcritical, or pitchfork bifurcation.

For the fixed points $X_{2,3}$ we have $f'(x = \pm\sqrt{\mu + 1}) = 3(\mu+1) - \mu = 3 + 2\mu$, and thus they are stable if $|3 + 2\mu| < 1$, which gives $-2 < \mu < -1$. For $\mu = -1$, we have $X_1 = X_2 = X_3$—this is a pitchfork bifurcation. Also as for $|\mu| < 1$, X_1 is stable—this is a subcritical pitchfork, and therefore $X_{2,3}$ are both unstable. Bifurcation diagram is shown in Figure 2.17.

Solution to Problem 2.4.

(i) *Fixed points* are the roots of the algebraic equation $x = f(x)$ or $x = 1 - ax^2$, which are

$$X_{1,2} = \frac{-1 \pm \sqrt{1 + 4a}}{2a}.$$

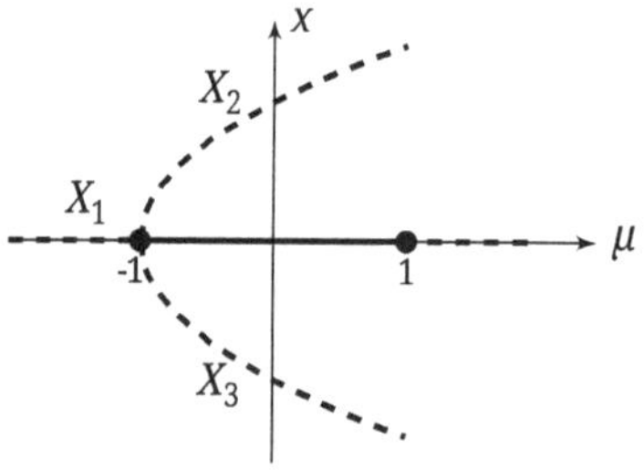

Figure 2.17: Bifurcation diagram for the map (2.17). Two bifurcation points are marked by filled circles: a subcritical pitchfork bifurcation at $(\mu, x) = (-1, 0)$ and a flip bifurcation at $(\mu, x) = (1, 0)$.

They exist when $a \geq -1/4$; at $a = -1/4$ we have a tangent bifurcation.
Stability of fixed points is determined by the value of the derivative $f'(x) = -2ax$ at the fixed point, which must lie within the interval $(-1, 1)$ for the fixed point to be stable.

$$f'(X_{1,2}) = -2a\frac{-1 \pm \sqrt{1+4a}}{2a} = 1 \mp \sqrt{1+4a}.$$

This derivative is always grater than 1 for $X_1 = \frac{-1-\sqrt{1+4a}}{2a}$. As for $X_2 = \frac{-1+\sqrt{1+4a}}{2a}$, it is stable when

$$-1 < 1 - \sqrt{1+4a} < 1 \quad \Rightarrow \quad -1 < \sqrt{1+4a} - 1 < 1 \quad \Rightarrow$$
$$0 < \sqrt{1+4a} < 2 \quad \Rightarrow \quad 0 < 1 + 4a < 4 \quad \Rightarrow \quad -1 < 4a < 3 \quad \Rightarrow \quad -1/4 < a < 3/4.$$

Thus, $X_2 = \frac{-1+\sqrt{1+4a}}{2a}$ is stable when $-\frac{1}{4} < a < \frac{3}{4}$. When $a = \frac{3}{4}$, $f'(X_2) = -1$, indicating a flip bifurcation at this parameter value.

(ii) *Two-periodic solution* is given by the fixed points of the two-fold composite map and is found as the roots of the following equation:

$$x = F(x) = f(f(x)) = 1 - a(1 - ax^2)^2.$$

Expanding the brackets, we obtain a fourth-degree polynomial

$$-a^3x^4 + 2a^2x^2 - x - a + 1 = 0.$$

This can be factorized, taking into account that the roots of $x = f(x)$ are also roots of the polynomial

$$-a^3x^4 + 2a^2x^2 - x - a + 1 = (ax^2 + x - 1)(\text{quadratic}) = 0.$$

The quadratic expression in the second bracket is found to be $-a^2x^2 + ax + a - 1$, with the roots

$$X_{3,4} = \frac{1 \pm \sqrt{4a-3}}{2a}.$$

These roots exist when $a > 3/4$. Thus, at $a = 3/4$ we observe a *flip bifurcation* for the map $x_{n+1} = 1 - ax^2$, and a *supercritical pitchfork bifurcation* for the composite map $x_{n+2} = -a^3x_n^4 + 2a^2x_n^2 - x_n - a + 1$.
This two-periodic solution is *stable* if $|F'(X_3)| < 1$ (note that $F'(X_3) = F'(X_4)$):

$$F'(X_3) = f'(X_3) \cdot f'(X_4) = (-2aX_3) \cdot (-2aX_4) = 4(1-a).$$

Thus, it is stable if

$$\left|4(1-a)\right| < 1 \quad \Rightarrow \quad -1/4 < a - 1 < 1/4 \quad \Rightarrow \quad 3/4 < a < 5/4.$$

At $a = \frac{5}{4}$, we observe the next round of period doubling.

Solution to Problem 2.5. The logistic map attractors corresponding to the numerical estimates of superstable points given in this question are shown in Figure 2.7. By gradually increasing the parameter r, we transition from a stable fixed point to a stable 2-cycle at the first bifurcation, followed by a 4-cycle, 8-cycle, 16-cycle, and so on (2^n-cycles). As r increases, successive *period-doubling bifurcations* occur, during which asymptotic stability is transferred from lower- to higher-order cycles (a process termed a *supercritical flip bifurcation*).

Feigenbaum observed that the period-doubling sequence proceeds broadly in geometric fashion in the limit of high-order cycles. The vertical distances are rescaled by the Feigenbaum constant $\alpha = 2.5029078\ldots$, and the horizontal distances by $\delta = 4.6692016\ldots$. In other words, the period doubling becomes self-similar as n increases, with r being reduced by the following factor of δ per doubling:

$$\delta = \lim_{n\to\infty} \delta_n, \quad \text{where } \delta_n = \frac{r\left(2^n\right) - r\left(2^{n-1}\right)}{r\left(2^{n+1}\right) - r\left(2^n\right)}, \tag{2.18}$$

and $r(2^n)$ is the r-value associated with the superstable fixed point of the 2^n-cycle.

For $n = 0$, a $2^0 = 1$-cycle is associated with $r = 2$.

For $n = 1$, a 2^1-cycle is associated with $r = 1 + \sqrt{5}$, and so on.

Evaluating (2.18) for $n = 1, 2, 3, 4, 5$ gives $\delta_n = 4.70894, 4.680771, 4.662958, 4.668408$ and 4.668968, which is consistent with convergence to the limit $\delta = 4.6692\ldots$, with five significant digits in the commonly quoted value.

Taking $\delta(n = 5) = 4.668968$ as our best estimate, and using the formula for a geometric progression, we can estimate the value of r corresponding to infinite periodicity

$$r(2^\infty) = \left[r(64) - r(32)\right] \cdot \frac{\delta}{\delta - 1} + r(32),$$

leading to the numerical estimate $r(2^\infty) \approx 3.569945$.

It is observed that around this value of parameter r, there is an accumulation point of the 2^∞-cycle on Figure 2.7: This marks the onset of chaos!

Solution to Problem 2.6. For the orbit x_i, $i = 0, 1, \dots, \infty$, the Lyapunov exponent is defined as

$$\lambda = \lim_{n\to\infty}\left(\frac{1}{n}\sum_{i=0}^{n-1}\ln|f'(x_i)|\right).$$

If $\lambda > 0$, then the system is sensitive to initial conditions, and if the orbits are bounded, the dynamics is chaotic.

For the given map, $f'(x) = 1$ for $0 \le x \le 0.5$, and $f'(x) = -2$ for $0.5 < x \le 1$.

At least every second iteration lies on the second branch (between 0.5 and 1); in fact, one can show that, on average, x_n falls on the second branch three times more often than on the first. Thus,

$$\lambda = \lim_{n\to\infty}\left(\frac{1}{n}\sum_{i=0}^{n-1}\ln|f'(x_i)|\right) > \frac{1}{2}(\ln 2 + \ln 1) = \frac{\ln 2}{2} > 0,$$

and therefore the behavior of the system is chaotic.

Solution to Problem 2.7.

(a) The derivative of the map is constant, and $f'(x) = 1$ for all values of x. Therefore, the Lyapunov exponent is

$$\lambda = \lim_{n\to\infty}\left(\frac{1}{n}\sum_{i=0}^{n-1}\ln|f'(x_i)|\right) = \frac{1}{n}n\ln 1 = \ln 1 = 0.$$

As the additive constant $\pi/4$ is irrational, the orbit densely fills the interval $[0, 1]$, but without sensitivity to initial conditions. Thus, we have quasiperiodic dynamics.

(b) The derivative $f'(x) = \sqrt{3}$ for all values of x (except for a single point $x = \frac{1}{\sqrt{3}}$, where the derivative is undefined). Therefore, the Lyapunov exponent is

$$\lambda = \lim_{n\to\infty}\left(\frac{1}{n}\sum_{i=0}^{n-1}\ln|f'(x_i)|\right) = \frac{1}{n}n\ln\sqrt{3} = \frac{1}{2}\ln 3 > 0$$

As $x_n \in [0, 1)$, i. e., bounded, and the system is sensitive to initial conditions, the dynamics are chaotic.

(c) This is a nonlinear map. Its fixed points are found as roots of $x = x^{\frac{1}{4}}$, which are $X_1 = 0$ and $X_2 = 1$.
The stability of fixed points is defined by the derivative $f'(x) = \frac{1}{4}x^{-\frac{3}{4}}$.
$f'(0) = \infty$, and therefore $X_1 = 0$ is unstable.
$f'(1) = \frac{1}{4}$, and therefore $X_2 = 1$ is stable.
As we have the stable fixed point $X_2 = 1$, where $f'(x) = \frac{1}{4}$, for the Lyapunov exponent we get

$$\lambda = \lim_{n\to\infty}\left(\frac{1}{n}\sum_{i=0}^{n-1}\ln f'(1)\right) = \frac{1}{n}n\ln\frac{1}{4} = -\ln 4 < 0$$

i. e., we have regular behaviour with $x_n \to 1$.

3 One-variable continuous dynamical systems

Consider a system whose state is described by a single variable x and its dynamics or evolution over time is given by a function $x(t)$. If time is considered as a continuous variable, the dynamics of the system is described by the rate of change, dx/dt, which is commonly represented using Newton's notation for the derivative, $\dot{x} = dx/dt$. The rate of change $\dot{x}$ is determined from the analysis of the system's properties and is used to derive the differential equation describing its dynamics.

Recall the classification of first-order differential equations [3]:

- $F(\dot{x}, x, t) = 0$ the general representation of a first-order differential equation.
- $\dot{x} = f(x, t)$—a first-order differential equation in explicit form, i. e., the rate of change, $\dot{x}$, is given explicitly by a function of both the dependent variable x and the independent variable t.
- $\dot{x} = a(t)x + b(t)$—a linear first-order differential equation; linear means that the function $f(x, t)$ is a linear function of the dependent variable x.
- $\dot{x} = f(x)$—an autonomous first-order differential equation; autonomous means that the rate of change $\dot{x}$ does not depend explicitly on time.
- $\dot{x} = ax + b$—a linear autonomous first-order differential equation.

The solution of the differential equation

$$\dot{x} = f(x, t) \tag{3.1}$$

is a function $x = x(t)$ such that the following identity holds:

$$\frac{dx(t)}{dt} \equiv f(x(t), t).$$

Equation (3.1) can be presented graphically on the (t, x) plane as illustrated in Figure 3.1. It defines a so-called *direction field* represented by vectors indicating the slop of the solution in each point (t, x) with components $(\Delta t, f(x, t)\Delta t)$ proportional to the rate of change of dependent variable x. These vectors are tangential to solutions of Eq. (3.1), which are called *integral curves, trajectories*, or *orbits*, and also illustrated in Figure 3.1. Solutions of differential equations are represented by collections of integral curves, which typically form a pattern reminiscent of flows in a viscous liquid. For this reason, differential equations describing continuous dynamical systems are often called *flows*.

The equation (3.1) should be supplied with an initial condition

$$x(t = 0) = x_0 \tag{3.2}$$

to specify a unique solution. The task of finding the unique solution to (3.1), subject to the initial condition (3.2), is called the Cauchy problem. The solution to this problem can be regarded as a function of both time and the initial condition $x = x(t, x_0)$.

https://doi.org/10.1515/9783111348339-003

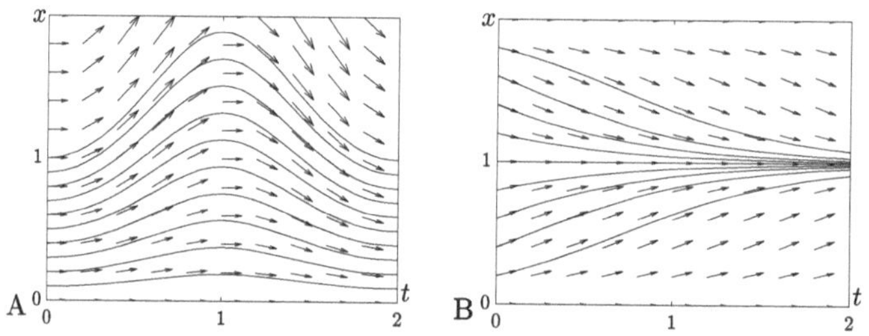

Figure 3.1: Direction fields and integral curves. Panel A: Direction field and integral curves for the nonautonomous equation $\dot{x} = x\sin(\pi t)$. **Panel B:** Direction field and integral curves for the autonomous equation $\dot{x} = 0.5\sin(\pi x)$.

If the state of the system is described by more than one variable, its dynamics is represented by the solution of a set (or system) of equations. A system of differential equations describes a so-called *continuous system*, as opposed to *discrete-time systems*, which are described by difference equations. We will consider systems of coupled differential equations in the next chapter.

3.1 Autonomous differential equations

Getting back to the equation (3.1), we note that

- $f(x,t) > 0$ means $\dot{x} > 0$, that is, x increases over time, and consequently the flow is directed along the positive x-axis.
- Similarly, $f(x,t) < 0$ means $\dot{x} < 0$, so x decreases over time, and the flow is directed along the negative x-axis.

Let us consider an autonomous differential equation

$$\dot{x} = f(x), \tag{3.3}$$

defined on a domain $x \in D$ (which may coincide with the entire real axis), where the function $f(x)$ has a continuous derivative, i. e., $f(x) \in C^1$. An important property of solutions to the autonomous equation (3.3) is that they are always monotonic (either nondecreasing or nonincreasing). Indeed, the slope of a solution is uniquely determined by the value of x, and a nonmonotonic solution would require the same value of x to correspond to different slopes, which is impossible.

The same argument implies that different solutions cannot cross each other. If two solutions coincide at some moment in time, say $x_1(t') = x_2(t')$, then they must coincide for all times, i. e., $x_1(t) \equiv x_2(t)$. More generally, if the function $f(x,t)$ in Eq. (3.1) is differentiable and its partial derivatives with respect to both arguments are continuous, i. e., $f(x,t) \in C^1$, then its solutions also cannot cross or touch each other. More about this statement can be found in [3].

Definition 3.1. A collection of all orbits of Eq. (3.1) presented on the plane (t, x) is called a **phase portrait**. Furthermore, for autonomous Eq. (3.3), the x-axis with arrows indicating direction of flows is called the **phase line** (illustrated in Figure 3.2).

As we noted before, $f(x) > 0$ in Eq. (3.3) means that x is increasing over time, while $f(x) < 0$ means it is decreasing. Important solutions are associated with the case when $f(x) = 0$, meaning that $\dot{x} = 0$ and x does not change over time.

Definition 3.2. Solutions such that $x(t) = c$ is constant are called **stationary solutions**. Other terms used to refer to such solutions are **equilibrium points** and **fixed points**.

Equilibrium points are the roots of the algebraic equation $f(x) = 0$. Since trajectories cannot cross each other, they also cannot cross equilibrium points. Therefore, equilibria divide the entire range of variable x into domains, such that in each domain the solutions behave similarly. The function $f(x)$ is equal to zero at equilibrium points and has a certain sign between neighboring equilibria.

If X_1 and X_2 are neighboring roots of $f(x) = 0$, then the solution $x(t)$ corresponding to the initial value $x_0 \in (X_1, X_2)$ is either an increasing or a decreasing function of time (depending on the sign of $f(x)$ between the equilibrium points X_1 and X_2). The same holds true for the domains between the highest equilibrium point and the upper boundary of the domain D and the lowest equilibrium point and the lower boundary of D. This is illustrated in Figure 3.2.

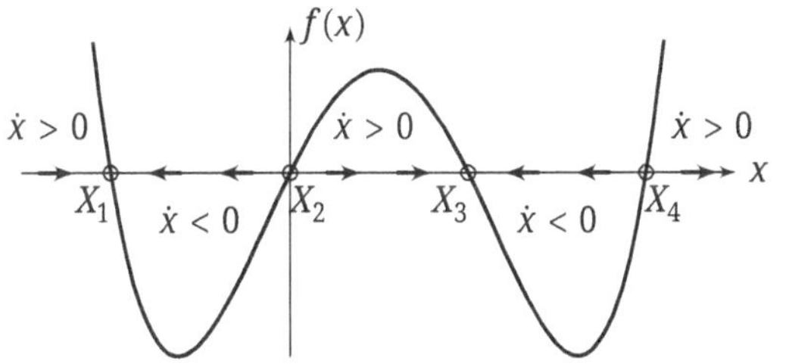

Figure 3.2: Graphical analysis of the equation $\dot{x} = f(x)$. The function $f(x)$ is a fourth-degree polynomial with four roots, X_1, X_2, X_3, and X_4, which correspond to the equilibrium points of the system. Solutions satisfy $x(t)$ increasing where $f(x) > 0$ and decreasing where $f(x) < 0$. The equilibria X_2 and X_4 are unstable since nearby trajectories diverge from them, whereas X_1 and X_3 are stable, with nearby trajectories converging to these points.

Furthermore, as solutions are strictly monotonic in each such domain and cannot cross its boundaries, it follows that they must tend towards one of the boundaries of the domain, that is, either to an equilibrium point (illustrated in Figure 3.1B) or to infinity. This brings us to the concept of stability of equilibria.

The stability of equilibrium points for flows is defined in the same way as the stability of fixed points for maps: There should exist a neighborhood of the equilibrium point such that orbits originating in this neighborhood do not diverge from the equilibrium (or converge to it, in the case of asymptotic stability). More formally, we state the following.

Definition 3.3. If X is an equilibrium point of Eq. (3.3), and there exist ε such that $\forall x_0$, defined by (3.2), and satisfying $|x_0 - X| < \varepsilon$, the solution $x(t, x_0) \to X$ as $t \to \infty$, then X is an asymptotically stable equilibrium point.

Thus, each solution of the flow (3.3) either converges to one of stable equilibrium points or diverge to infinity. Furthermore, all solutions originating in a domain bounded by two neighboring equilibrium points converge to one of them, i. e. they exhibit qualitatively the same behaviour. It also follows that solutions located in two neighboring domains can converge to the same equilibrium. This leads us to the consideration of larger domains in which solutions originating within them converge to the same equilibrium.

Definition 3.4. Let X be a stable equilibrium point of the flow (3.3). The collection of all values x_0 such that $x(t, x_0) \to X$ is called the **basin of attraction** of the equilibrium point X.

The analysis of differential equations based on the study of their equilibria and focused on long-range behaviour of solutions with different initial conditions is called *qualitative analysis.*

3.2 Stability of equilibrium points

As is evident from Figure 3.2, the plot of the function $f(x)$ allows us not only to identify equilibria (roots of $f(x) = 0$), but also to determine their stability. A fixed point is stable if the flow to its left is directed to the right (i. e., $\dot{x} > 0$) and the flow to its right is directed to the left (i. e., $\dot{x} < 0$). Thus, if $f(x)$ changes sign from positive to negative at a fixed point, the point is stable; otherwise, it is unstable. From this observation, it is easy to conclude that the derivative $f'(x)$ is negative at a stable fixed point and positive at an unstable one. In this section, we will prove this statement analytically; but before doing so, we will consider a very important example of a flow defined by a linear function.

Example 3.1. Analyze how the orbits of the linear equation $\dot{x} = ax + b$ depend on the parameters a and b.

Solution:
Let us first consider the case when $b = 0$ and find its analytical solution. Separate the variables as follows:

$$\dot{x} = ax \quad \Rightarrow \quad \frac{dx}{dt} = ax \quad \Rightarrow \quad \frac{dx}{x} = adt.$$

Integrate as follows:

$$\int \frac{dx}{x} = \int a dt \quad \Rightarrow \quad \ln|x| = at + C \quad \Rightarrow \quad x = e^C e^{at} = Ae^{at}.$$

If $x(t = 0) = x_0$, then

$$x_0 = Ae^{a \cdot 0} = A \quad \Rightarrow \quad A = x_0.$$

Thus, the solution is

$$x(t) = x_0 e^{at}.$$

From the solution, we conclude that
- If $a < 0$, then $x(t) \to 0$ as $t \to \infty$.
- If $a > 0$, then $x(t) \to \pm\infty$ as $t \to \infty$.

In terms of qualitative analysis: The considered equation with $b = 0$ has a single fixed point, $X = 0$, which is stable when $a < 0$ and unstable when $a > 0$. This is illustrated by integral curves for different values of x_0 converging to $X = 0$ in the case of $a < 0$ (shown in Figure 3.3A) and diverging from $X = 0$ in the case of $a > 0$ (Figure 3.3B). The x-axis, with arrows indicating the direction of flows, represents the phase line.

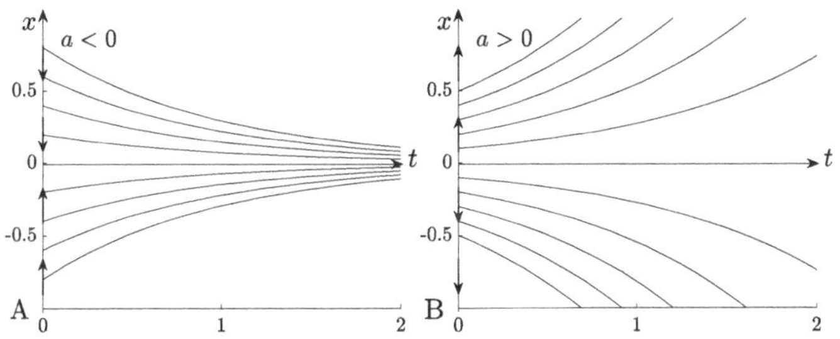

Figure 3.3: Solutions of the linear equation $\dot{x} = ax$. Panel **A** shows solutions for $a < 0$ (here $a = -1$), where all trajectories converge to the equilibrium $X = 0$, indicating that this fixed point is stable with the entire x-axis as its domain of attraction. Panel **B** shows solutions for $a > 0$ (here $a = 1$), where trajectories diverge from $X = 0$, demonstrating that the equilibrium is unstable. In both panels, the phase line is represented by the x-axis, with arrows indicating the direction of the flow.

When $a = 0$, we have neither divergence nor convergence to the fixed point $X = 0$, and this case is referred to as *neutral stability* of the fixed point.

The case when $b \neq 0$ is not much different. The fixed point is now found as the root of $ax + b = 0$, i. e. $X = -b/a$. Following the same steps as before, we can show that the analytical solution in this case is

$$x(t) = (x_0 - X)e^{at} + X.$$

It is evident from this solution that the fixed point is again stable if $a < 0$ and unstable if $a > 0$. Note that the derivative of the map-generating function $f(x) = ax$ is $f'(x) = a$ for all values of x, including the fixed point X. Thus, we can state the condition for the stability of the fixed point as follows: The only fixed point of the linear map is stable if the derivative of the map-generating function is negative and unstable if it is positive.

Theorem 3.1. *An equilibrium point X of the autonomous Eq.* (3.3) *is asymptotically stable iff $f'(X) < 0$ and unstable iff $f'(X) > 0$.*

Proof. Make the substitution

$$x(t) = X + \varepsilon(t),$$

where $\varepsilon(t)$ is a new variable representing the difference between $x(t)$ and X. If $x_0 \approx X$, then $\varepsilon_0 = \varepsilon(t = 0)$ is small, $|\varepsilon_0| \ll 1$, and we can use the linear approximation of Eq. (3.3). Thus, we get

$$\dot{x}(t) = \frac{d}{dt}(X + \varepsilon) = \frac{d}{dt}(\varepsilon) = \dot{\varepsilon} = f(x) = f(X + \varepsilon) \approx f(X) + f'(X)\varepsilon.$$

$f(X) = 0$ since X is a stationary point, and therefore,

$$f(X) + f'(X)\varepsilon = f'(X)\varepsilon.$$

Thus, the first-order approximation gives:

$$\dot{\varepsilon} = f'(X)\varepsilon \quad \Rightarrow \quad \varepsilon(t) = \varepsilon_0 e^{f'(X)t}.$$

Using the above result for a linear differential equation, we conclude that as $t \to \infty$, $\varepsilon(t)$ tends to zero if $f'(X)$ is negative and tends to infinity if $f'(X)$ is positive. Furthermore, $\varepsilon(t) \to 0$ means that $x(t) \to X$, and X is a stable fixed point.

Thus, if $f'(X) < 0$, then X is asymptotically stable; and if $f'(X) > 0$, then X is unstable. □

Example 3.2. Consider the autonomous differential equation

$$\dot{x} = f(x) - x,$$

where $f(x)$ is a continuous and differentiable function with domain $[0, 1]$ and range $[0, 1]$. $f(x) - x$ has three fixed points: two stable fixed points at $x = 0$ and $x = 1$ and an unstable fixed point at $x = 0.5$. Sketch $f(x)$ and find the basin of attraction of each stable fixed point.

Solution:
$f(x)$ crosses the line $y = x$ at fixed points $X_1 = 0$, $X_2 = 1/2$ and $X_3 = 1$. Stable fixed points are such that $f'(x) < 1$, whereas for unstable fixed points $f'(x) > 1$.

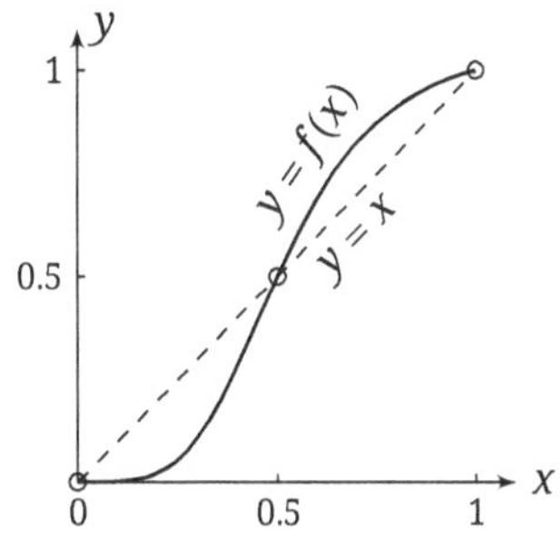

Figure 3.4: Sketch of the curve $f(x)$ in Example 3.2.

The simplest (monotonically increasing) $f(x)$ satisfying all given conditions (differentiable, with a range [0,1], and crossing the diagonal line in the given three points, so that its derivative is less than 1 at $x = 0$ and $x = 1$), is shown on Figure 3.4. $f(x)$ is represented in this figure by the cubic polynomial $f(x) = -x(x^2 - 1.5x - 0.5)$, giving $\dot{x} = x(x - 0.5)(1 - x)$.

When the initial value $x(t = 0) = x_0$ is lower than 1/2, $x(t)$ converges to the fixed point $X_1 = 0$. When the value x_0 is greater that 1/2, $x(t)$ converges to the fixed point $X_3 = 1$. Thus, $0 \leq x < 0.5$ is the basin of attraction for $X_1 = 0$ and $0.5 < x \leq 1$ for $X_3 = 1$.

3.3 Logistic equation

In the chapter devoted to one-dimensional maps, we placed significant emphasis on the study of the logistic map. We will now consider its continuous counterpart, known as the logistic equation.

3.3.1 Verhulst model

The logistic equation originated from studies of population dynamics. The simplest population growth model, known as the Malthusian model, was introduced by the English scientist Thomas Robert Malthus in 1798. It is based on the assumption that the rate of change in population size is proportional to the size itself

$$\dot{N} = rN,$$

where the variable N represents the size of the population and the parameter r is the so-called intrinsic reproduction rate. Consequently, the model is considered for $N \geq 0$ and usually for $r > 0$ (growing population).

The Malthusian model satisfactorily reproduced the exponential growth of the human population in England during the 16th and 17th centuries; however, in the long run, it predicts infinite growth of the population size, which is unrealistic. The model was modified by the Belgian scientist Pierre François Verhulst in 1838, who assumed that the intrinsic reproduction rate decreases with an increase in the population size. The Verhulst model is given by the equation

$$\dot{N} = rN\left(1 - \frac{N}{K}\right), \tag{3.4}$$

where the new parameter K is called the carrying capacity. According to this equation, in the long run the population size saturates at $N = K$.

3.3.2 Nondimensionalization

The Verhulst model has two parameters, which, however, can be omitted using the so-called nondimensionalization of the equation. In the Verhulst model, we have two variables: the independent variable t, which has the dimension of time, and the dependent variable N, which has the dimension of population size (i. e., number of organisms). There are also two dimensional parameters r and K.

The expression in the brackets should be dimensionless (since the first term is nondimensional), and therefore the dimension of K is the same as that of N. Thus, we can introduce the nondimensional variable $x = N/K$ and rewrite Eq. (3.4) as

$$\dot{x} = rx(1 - x). \tag{3.5}$$

Now, $x(1 - x)$ is nondimensional, and therefore $\dot{x}/r$ is also nondimensional

$$\frac{\dot{x}}{r} = \frac{dx}{rdt} = \frac{dx}{d\tilde{t}},$$

where $\tilde{t} = rt$ is a nondimensional substitution for t.

Thus, we arrive at the nondimensional counterpart of Eq. (3.4), which is given by

$$\dot{x} = x(1 - x), \tag{3.6}$$

where $\dot{x}$ denotes the derivative with respect to the nondimensional variable $\tilde{t}$. Alternatively, we can consider this equation as Eq. (3.5), where $r = 1$ is fixed, in which case $\tilde{t} = t$. The term *logistic equation* applies to all three differential equations (3.4)–(3.6). In what follows, we will focus on Eq. (3.6) and treat it as the mathematically appropriate form of the logistic equation.

3.3.3 Analysis of the logistic equation

Equation (3.6) can be integrated analytically using separation of variables

$$\dot{x} = \frac{dx}{dt} = x(1 - x) \quad \Rightarrow \quad \int \frac{dx}{x(1 - x)} = \int dt.$$

The left-hand side can be decomposed into partial fractions

$$\int \frac{dx}{x(1-x)} = \int \left(\frac{1}{x} - \frac{1}{x-1}\right) dx = \ln|x| - \ln|x-1| = \ln\left|\frac{x}{x-1}\right|.$$

So,

$$\int \frac{dx}{x(1-x)} = \int dt \quad \Rightarrow \quad \ln\left|\frac{x}{x-1}\right| = t + c.$$

Exponentiate both sides of the last equation

$$\left|\frac{x}{x-1}\right| = e^{t+c} \quad \Rightarrow \quad \frac{x}{x-1} = Ae^t,$$

where A is a constant determined by the initial condition. Solving for x, we get

$$x = \frac{Ae^t}{Ae^t - 1} = \frac{1}{1 - \frac{1}{Ae^t}} = \frac{1}{1 - \frac{e^{-t}}{A}} = \frac{1}{1 - ce^{-t}},$$

where $c = 1/A$. Using the initial condition $x(t = 0) = x_0$, we find c as follows:

$$x_0 = \frac{1}{1-c} \quad \Rightarrow \quad x_0(1-c) = 1 \quad \Rightarrow \quad x_0 - 1 = cx_0 \quad \Rightarrow \quad c = \frac{x_0 - 1}{x_0}.$$

Therefore, the solution is given by

$$x(t) = \frac{1}{1 - \frac{x_0 - 1}{x_0} e^{-t}}. \tag{3.7}$$

Now, since $e^{-t} \to 0$ as $t \to \infty$, it follows that $x(t) \to 1$ as $t \to \infty$ for any initial value x_0.

Using qualitative analysis of Eq. (3.6), we find that there are two fixed points, given by the roots of $f(x) = x(1-x) = 0$, namely, $X_1 = 0$ and $X_2 = 1$ (see Figure 3.5A). The stability of the fixed points is determined by the derivative $f'(x) = 1 - 2x$. Evaluating this at the fixed points, we find $f'(0) = 1 > 0$, and therefore $X_1 = 0$ is unstable, while $f'(1) = -1 < 0$, and therefore $X_2 = 1$ is stable.

From the qualitative analysis, it appears that $x \to -\infty$ if $x_0 < 0$ (Figure 3.5A). This seems to contradict to solution found analytically (3.7), which implies that x tends to 1 for any x_0. To resolve this contradiction, we plotted several solutions, including one corresponding to $x_0 < 0$ (see Figure 3.5B). The contradiction is resolved when we note that for $x_0 < 0$, the solution $x(t)$ indeed tends to $-\infty$ at a finite time t^*, *as predicted by the qualitative analysis*. At $t = t^*$ the solution changes sign; for $t > t^*$ it is positive and converges to 1 as $t \to \infty$, in agreement with the analytical result (3.7).

The time t^* at which the solution reaches $-\infty$—called the *singularity point*—can be found from the solution formula (3.7) as the time when the denominator becomes zero

$$t^* = \ln\left(1 - \frac{1}{x_0}\right).$$

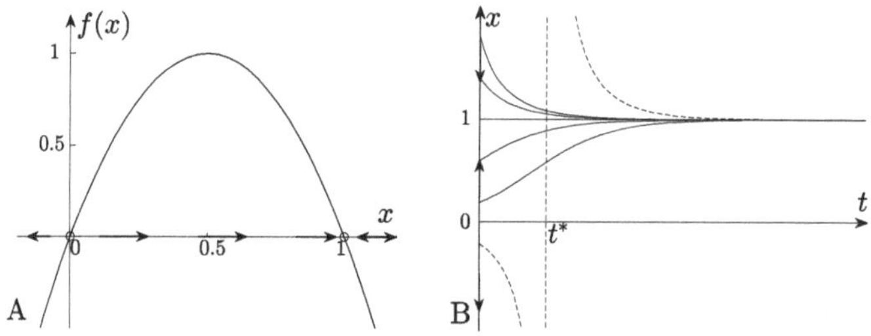

Figure 3.5: Solutions of the logistic equation. Panel **A** shows the graph of $f(x) = x(1-x)$, which intersects the x-axis at $X_1 = 0$ and $X_2 = 1$, the unstable and stable equilibria of the logistic equation, respectively. Panel **B** displays solutions starting from initial conditions $x_0 = -0.2, 0.2, 0.6, 1.0, 1.4$, and 1.8. For $x_0 > 0$, trajectories converge to $x = 1$ as $t \to \infty$ (solid curves). For $x_0 < 0$, the solution decreases, reaches $-\infty$ at a finite time $t = t^*$, changes sign, and subsequently converges to $x = 1$ (dashed curve). In both panels, the phase line is represented by the x-axis with arrows indicating the direction of the flow.

3.4 Bifurcation theory (equations with parameters)

As we saw in the previous chapter, the number, location and stability of fixed points in maps with a parameter depend on its value. This is vividly manifested in the period-doubling cascade, when the number of fixed points keeps increasing as the value of the parameter r increases in the logistic map. The change in the number and stability of fixed points is associated with bifurcations taking place at certain parameter values. In this section, we will introduce a formal definition of bifurcation, which will later be used for the analysis of bifurcations occurring in flows with a parameter.

Example 3.3. The population dynamics of a species of animal subject to hunting at a constant rate E is modelled by

$$\frac{dN}{dt} = rN\left(1 - \frac{N}{K}\right) - E,$$

which is a Verhulst model (3.4) with an extra term $E > 0$ representing the hunting quota. Using the substitutions $n = N/K$ and $\tau = rt$, the equation is nondimensionalized and transforms into

$$\frac{dn}{d\tau} = \dot{n} = n(1-n) - h, \tag{3.8}$$

with the single parameter $h = E/(Kr)$. Equilibria are roots of the algebraic equation $n(1-n) - h = 0$ and depend on the parameter h

$$N_{1,2} = \frac{1 \pm \sqrt{1-4h}}{2}.$$

They exist if $h < 0.25$. To analyze the stability of the equilibria, we check the sign of the derivative $f'(n) = 1 - 2n$ and find that it is negative for $N_2 = \frac{1+\sqrt{1-4h}}{2}$ and positive for

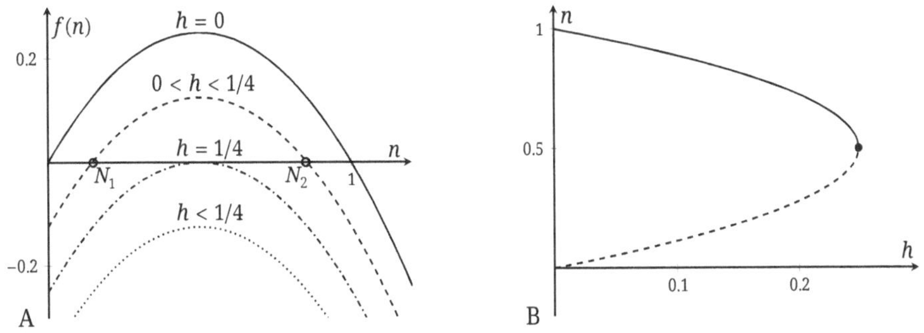

Figure 3.6: Dependence of equilibria on the parameter in Example 3.3. Panel A: Plots of $f(n) = n(1 - n) - h$ for $h = 0, 1/8, 1/4$, and $3/8$. As h increases, the root N_1 ($N_1 = 0$ for $h = 0$) increases and the root N_2 ($N_2 = 1$ for $h = 0$) decreases. The two roots coalesce at $N = 1/2$ when $h = 1/4$ and disappear for $h > 1/4$. **Panel B:** Bifurcation diagram for Eq. (3.8). The point $(h, n) = (0.25, 0.5)$ (marked by a solid circle) is a *bifurcation point*, where a *fold bifurcation* occurs. For $h < 0.25$, two fixed points exist; for $h > 0.25$, none remain. The equilibrium on the upper branch (N_2) is stable (solid line), while that on the lower branch (N_1) is unstable (dashed line).

$N_1 = \frac{1-\sqrt{1-4h}}{2}$. Thus, N_2 is stable and N_1 is unstable. With an increase in h, N_2 decreases and N_1 increases (see Figure 3.6A), and they collide when $h = 0.25$. At this value of the parameter h, we have a so-called *fold bifurcation*. The bifurcation diagram for Eq. (3.8) is shown in Figure 3.6B.

We begin with some preliminary definitions before introducing a formal definition of bifurcation.

Definition 3.5. Two equations $\dot{x} = f(x)$ and $\dot{y} = g(y)$ are said to be **topologically equivalent** if there exists a *smooth* transformation given by a continuous, bijective function $x = h(y)$ with a continuous inverse, such that all orbits of $\dot{x} = f(x)$ are mapped to the orbits of $\dot{y} = g(y)$, preserving the direction of the dynamics.

In other words, two flows are topologically equivalent if the number of fixed points and the stability of the corresponding fixed points (when both sets of fixed points are, for example, ordered increasingly) are the same for both flows.

Definition 3.6. A flow with parameter $\dot{x} = f(x, c)$ is said to be **structurally stable** at $c = C$ if the flow $\dot{x} = f(x, c)$ is topologically equivalent to $\dot{x} = f(x, C)$ for all values of c in a neighborhood of C.

More formally, a flow $\dot{x} = f(x, c)$ is structurally stable at $c = C$ if $\exists \varepsilon > 0$ such that $\forall c$ satisfying $|c - C| < \varepsilon$, the flows $\dot{x} = f(x, c)$, and $\dot{x} = f(x, C)$ are topologically equivalent.

Definition 3.7. Consider the parameter-dependent equation $\dot{x} = f(x, c)$. A value $c = C$ is called a **bifurcation value** if the flow $\dot{x} = f(x, C)$ is not structurally stable.

Let $\dot{x} = f(x, c)$ have an equilibrium point at (X, C), i. e., $f(X, C) = 0$. We can perform a change of variables to shift both the equilibrium point and the parameter value so that the transformed system has an equilibrium at $(0, 0)$.

Specifically, define new variables $y = x - X$ and $a = c - C$. Then, we have

$$\dot{x} = \frac{dx}{dt} = \frac{d(y + X)}{dt} = \frac{dy}{dt} = \dot{y} = f(x, c) = f(y + X, a + C) = F(y, a).$$

Now, $F(0, 0) = f(X, C) = 0$, so $(0, 0)$ is an equilibrium point of the transformed system $\dot{y} = F(y, a)$. Thus, in the following theorem we assume, without loss of generality, that the flow has an equilibrium at $(0, 0)$.

Theorem 3.2. *If $\dot{x} = f(x, c)$ has an equilibrium point at $(0, 0)$ and $f_x(0, 0) \neq 0$ (where $f_x(x, c)$ denotes the partial derivative of $f(x, c)$ with respect to x), then there exists a neighborhood of $c = 0$ in which the system has a unique equilibrium point whose stability is the same as that at $(0, 0)$.*

Proof. Considering $|c| \ll 1$ we can use a linear approximation of $f(x, c)$ around $(0, 0)$

$$\dot{x} = f(x, c) \approx f(0, 0) + f_x(0, 0)x + f_c(0, 0)c = f_x(0, 0)x + f_c(0, 0)c \tag{3.9}$$

since $f(0, 0) = 0$ as $(0, 0)$ is an equilibrium point.

To find the dependence of the equilibrium point, X, on parameter c, set $\dot{x} = 0$

$$f_x(0, 0)X + f_c(0, 0)c = 0 \quad \Rightarrow \quad X(c) = -\frac{f_c(0, 0)}{f_x(0, 0)}c. \tag{3.10}$$

Thus, a small variation in parameter c, $|c| \ll 1$, results in a small shift of the equilibrium point $X(c) = O(c)$.

The stability of this equilibrium is determined by the sign of $f_x(X(c), c)$. By continuity of f_x, if $f_x(0, 0) < 0$, then $\exists \varepsilon$ such that if $|c| < \varepsilon$, then $f_x(x, c) < 0$, i. e., the equilibrium remains stable in a neighborhood of $(0, 0)$. Similarly, if $f_x(0, 0) > 0$, then the equilibrium is unstable for sufficiently small $|c|$.

Thus, a sufficiently small variation of the parameter c results in a small shift of the equilibrium from $(0, 0)$ given by equation (3.10), without altering its stability (see Figure 3.7). □

Definition 3.8. An equilibrium point X of the flow $\dot{x} = f(x)$ is called **hyperbolic** if $f'(X) \neq 0$.

Theorem 3.3. *If the flow $\dot{x} = f(x, c)$ has a finite number of equilibria at $c = 0$ and all of them are hyperbolic, then the flow is structurally stable at $c = 0$.*

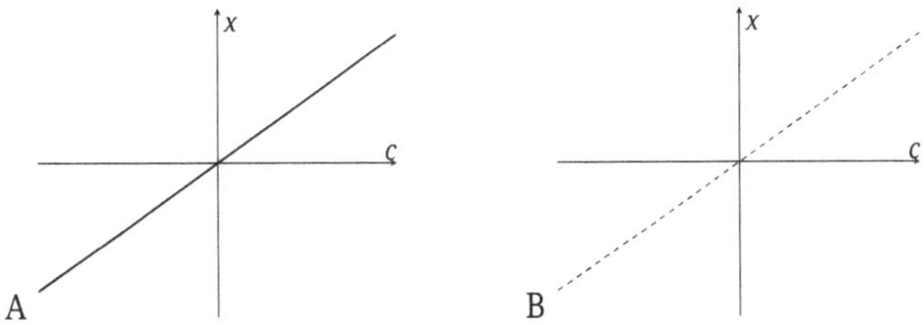

Figure 3.7: On structural stability and hyperbolic equilibria. Panel A: The stable equilibrium at $(0,0)$ of the linear flow (3.9) (where $f_x(0,0) < 0$) shifts along a straight line given by (3.10), but remains stable as the parameter c varies near $c = 0$. **Panel B:** The unstable equilibrium at $(0,0)$ (where $f_x(0,0) > 0$) also shifts along a straight line, yet remains unstable for c in a neighborhood of 0.

Proof. Let n be the number of equilibria of the flow $\dot{x} = f(x,c)$ at $c = 0$. Consider the i-th equilibrium point. By Theorem 3.2, for each such equilibrium, there exists $\varepsilon_i > 0$ such that for all c satisfying $|c| < \varepsilon_i$, the stability of the i-th equilibrium remains unchanged.

Define $\varepsilon = \min\{\varepsilon_1, \varepsilon_2, \dots, \varepsilon_n\}$. Then, for all $|c| < \varepsilon$, each equilibrium persists with the same stability, and the number of equilibria remains the same. Hence, the flow $\dot{x} = f(x,c)$ is structurally stable at $c = 0$. □

3.5 Classification of bifurcations

Consider the flow $\dot{x} = f(x,c)$. Let $(x,c) = (0,0)$ be an isolated nonhyperbolic equilibrium point of this flow. This means that $f(0,0) = 0$ (it is an equilibrium point) and $f_x(0,0) = 0$ (it is nonhyperbolic). Consequently, the linear approximation of the flow at $(x,c) = (0,0)$ reduces to the single term

$$\dot{x} = f(x,c) \approx f(0,0) + f_x(0,0)x + f_c(0,0)c = f_c(0,0)c. \tag{3.11}$$

According to this approximation, the flow has no fixed points when $c \neq 0$ and infinitely many fixed points (any $x \in \mathbb{R}$) when $c = 0$. Clearly, this does not reflect the actual behavior of the considered flow, and therefore its linear approximation turns out to be inadequate.

Therefore, to capture the essential properties of the flow, we need to consider its quadratic (second-order Taylor) approximation around the nonhyperbolic equilibrium point $(0,0)$, which is given by

$$\dot{x} = f(x,c) \approx f_c(0,0)c + \frac{1}{2}f_{xx}(0,0)x^2 + f_{xc}(0,0)xc + \frac{1}{2}f_{cc}(0,0)c^2. \tag{3.12}$$

The classification of bifurcations that can occur at nonhyperbolic equilibria is based on the analysis of the second-order approximation given in Eq. (3.12), together with the third-order approximation, which we present shortly.

3.5.1 Fold bifurcation

The analysis of Eq. (3.12) is based on comparing the relative magnitudes of the terms on its right-hand side. If $f_c(0,0) \neq 0$, then the linear term $f_c(0,0)c$ dominates over the three quadratic terms. However, we cannot disregard all three second-order terms, as doing so would lead to the inadequate linear approximation given by Eq. (3.11). Therefore, at least one of the quadratic terms must be of the same order of magnitude as the linear term to ensure a meaningful approximation.

To present the analysis of Eq. (3.12) in a compact form, we introduce the following notations: $a_1 = f_c(0,0)$, $a_2 = \frac{1}{2}f_{xx}(0,0)$, $a_3 = f_{xc}(0,0)$, and $a_4 = \frac{1}{2}f_{cc}(0,0)$. With these notations, Eq. (3.12) can be rewritten as

$$\dot{x} = a_1 c + a_2 x^2 + a_3 xc + a_4 c^2. \tag{3.13}$$

We know that the approximation becomes more accurate as both the variable x and the parameter c tend to zero. Let us compare the quadratic terms with the linear term in this limit.

First, consider the term $a_4 c^2$

$$\lim_{x,c\to 0} \frac{a_4 c^2}{a_1 c} = \lim_{c\to 0} \frac{a_4}{a_1} c = 0 \quad \Rightarrow \quad a_4 c^2 = o(a_1 c),$$

which shows that $a_4 c^2$ is asymptotically smaller than the linear term and can therefore be neglected.

Similarly, for the mixed term $a_3 xc$ we have

$$\lim_{x,c\to 0} \frac{a_3 xc}{a_1 c} = \lim_{x\to 0} \frac{a_3}{a_1} x = 0 \quad \Rightarrow \quad a_3 xc = o(a_1 c),$$

indicating that $a_3 xc$ is also negligible compared to the linear term.

Thus, Eq. (3.13) simplifies to

$$\dot{x} = a_1 c + a_2 x^2, \tag{3.14}$$

which can further be reduced to the normal form of a *fold bifurcation* through an appropriate rescaling of variables.

To proceed with Eq. (3.14), we introduce a new variable $\tilde{x}$ and a new parameter $\tilde{c}$ through appropriate substitutions, which will differ slightly depending on the signs of a_1 and a_2. These substitutions will be chosen such that $\tilde{x}$ is an increasing function of x and $\tilde{c}$ is an increasing function of c, thereby preserving the orientation of the phase space in both the variable and the parameter directions.

Case 1. $a_1 > 0$ and $a_2 > 0$. We perform the substitution $\tilde{x} = a_2 x$ and $\tilde{c} = a_1 a_2 c$, which transforms Eq. (3.14) into

$$\frac{\dot{\tilde{x}}}{a_2} = a_1 \frac{\tilde{c}}{a_1 a_2} + a_2 \left(\frac{\tilde{x}}{a_2} \right)^2 = \frac{\tilde{c}}{a_2} + \frac{\tilde{x}^2}{a_2}.$$

Multiplying both sides by a_2 gives

$$\dot{\tilde{x}} = \tilde{c} + \tilde{x}^2.$$

After dropping the tildes (i. e., reverting to x and c), we obtain the *normal form* of the fold bifurcation

$$\dot{x} = c + x^2.$$

Equilibrium points are the roots of the equation $c + x^2 = 0$, which are given by

$$X_{1,2} = \pm\sqrt{-c},$$

and they exist only when $c < 0$.

To analyze the stability of these equilibria, we compute the derivative of $f(x,c) = c + x^2$ with respect to x

$$f_x(x,c) = 2x.$$

Evaluating f_x at $X_1 = \sqrt{-c}$ yields

$$f_x(\sqrt{-c}, c) = 2\sqrt{-c} > 0,$$

thus, the equilibrium $X_1 = \sqrt{-c}$ is unstable.

Similarly, evaluating at $X_2 = -\sqrt{-c}$ gives

$$f_x(-\sqrt{-c}, c) = -2\sqrt{-c} < 0,$$

therefore, the equilibrium $X_2 = -\sqrt{-c}$ is stable.

The bifurcation diagram for this scenario (Case 1) is illustrated in Figure 3.8A.

For arbitrary signs of a_1 and a_2 in Eq. (3.14), we perform the substitution $\tilde{x} = |a_2|x$ and $\tilde{c} = |a_1 a_2|c$, which transforms the equation into

$$\frac{\dot{\tilde{x}}}{|a_2|} = \frac{a_1 \tilde{c}}{|a_1 a_2|} + a_2 \frac{\tilde{x}^2}{|a_2|^2},$$

or equivalently,

$$\dot{\tilde{x}} = \frac{a_1}{|a_1|}\tilde{c} + \frac{a_2}{|a_2|}\tilde{x}^2.$$

Noting that $\frac{a_1}{|a_1|} = \pm 1$ and $\frac{a_2}{|a_2|} = \pm 1$, and reverting to the original notation (omitting tildes), we arrive at the equation

$$\dot{x} = \pm c \pm x^2. \tag{3.15}$$

This equation represents the *normal form* for a fold bifurcation and encompasses all four possible cases, corresponding to the combinations of signs of a_1 and a_2.

The conditions for a fold bifurcation to occur are

$$f_c(0,0) \neq 0 \quad \text{and} \quad f_{xx}(0,0) \neq 0,$$

which are known as the *nondegeneracy conditions* for a fold bifurcation. A characteristic feature of a fold bifurcation is that a pair of equilibria—one stable and one unstable—appear or disappear simultaneously at the bifurcation point.

The four cases of fold bifurcation correspond to the different combinations of signs in the right-hand side of Eq. (3.15). Case 1, where both signs are positive, was considered previously. The remaining cases are as follows.

Case 2. $\dot{x} = -c + x^2$.

The equilibrium points of this equation are given by $X_{1,2} = \pm\sqrt{c}$, which exist when $c > 0$. Regarding stability,

$$\begin{aligned} f_x(x = \sqrt{c}, c) &= 2\sqrt{c} > 0 \quad \Rightarrow \quad X_1 = \sqrt{c} \quad \text{is unstable.} \\ f_x(x = -\sqrt{c}, c) &= -2\sqrt{c} < 0 \quad \Rightarrow \quad X_2 = -\sqrt{c} \quad \text{is stable.} \end{aligned}$$

Case 3. $\dot{x} = f(x,c) = c - x^2$.

The equilibrium points are $X_{1,2} = \pm\sqrt{c}$ and exists if $c > 0$. Now $f_x = -2x$ and shall be evaluated at $\pm\sqrt{c}$.

$$\begin{aligned} f_x(x = \sqrt{c}, c) &= -2\sqrt{c} < 0 \quad \Rightarrow \quad X_1 = \sqrt{c} \quad \text{is stable} \\ f_x(x = -\sqrt{c}, c) &= 2\sqrt{c} > 0 \quad \Rightarrow \quad X_2 = -\sqrt{c} \quad \text{is unstable} \end{aligned}$$

Case 4. $\dot{x} = f(x,c) = -c - x^2$.

The equilibrium points of this equation are given by $X_{1,2} = \pm\sqrt{-c}$, which exists if $c < 0$. We shall evaluate f_x at $\pm\sqrt{-c}$.

$$\begin{aligned} f_x(x = \sqrt{-c}, c) &= -2\sqrt{-c} < 0 \quad \Rightarrow \quad X_1 = \sqrt{-c} \quad \text{is stable} \\ f_x(x = -\sqrt{-c}, c) &= 2\sqrt{-c} > 0 \quad \Rightarrow \quad X_2 = -\sqrt{-c} \quad \text{is unstable} \end{aligned}$$

Bifurcation diagrams for all four cases of the fold bifurcation are shown in Figure 3.8. In each case the bifurcation diagram is represented by a parabola, with one branch corresponding to a stable equilibrium point and the other to an unstable one.

Example 3.4. Analyze bifurcation taking place in the flow

$$\dot{x} = r + x - \ln(1 + x), \quad x > -1.$$

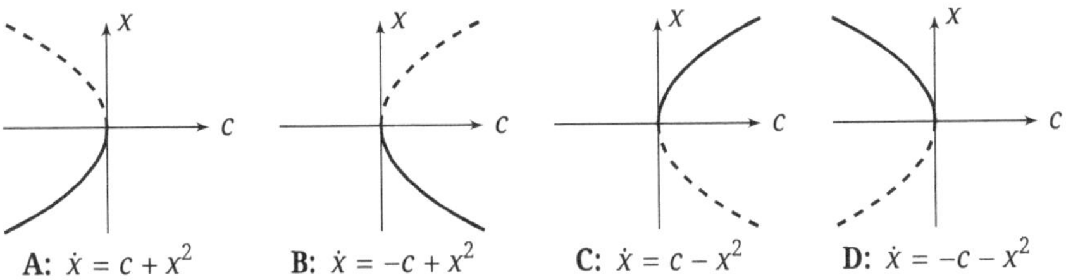

Figure 3.8: Bifurcation diagrams for the four cases of a fold bifurcation. Stable equilibria are shown with solid lines and unstable equilibria with dashed lines.

Solution:
Bifurcations take place at nonhyperbolic fixed points, that is, at points where $f(x, r) = 0$ and $f_x(x, r) = 0$. Let us start with the second condition

$$f_x'(x, r) = 0 \quad \Rightarrow \quad 1 - \frac{1}{1+x} = 0 \quad \Rightarrow \quad x = 0.$$

Thus, the bifurcation takes place at $x = 0$ for a certain value of the parameter r, which is found from the first condition $f(x = 0, r) = 0 \Rightarrow r - \ln 1 = 0 \Rightarrow r = 0$. Thus, the bifurcation point is $(x, r) = (0, 0)$.

To determine what kind of bifurcation takes place, we replace the logarithm with its approximation, which should be at least of the second order: $\ln(1 + x) \approx x - \frac{x^2}{2}$. The second-order approximation of the flow is

$$\dot{x} = r + x - x + \frac{x^2}{2} = r + \frac{x^2}{2}$$

which is in an almost normal form for a *fold bifurcation* (given by Eq. (3.15), Case 1).

There are no fixed points for $r > 0$, and two fixed points $X_{1,2} = \pm\sqrt{-2r}$ for $r < 0$. It is easy to check that the fixed point given by the positive number is unstable, while the other one is stable. The bifurcation diagram for this flow is of the type shown in Figure 3.8A.

3.5.2 Transcritical bifurcation

It was essential in our discussion of the fold bifurcation that $f_c(0, 0) \neq 0$ and $f_{xx}(0, 0) \neq 0$, that is, the *nondegeneracy conditions* for a fold bifurcation hold. If these conditions are violated, other types of bifurcations can occur.

Now, assume that in the quadratic approximation of the flow derived earlier in Eq. (3.13), the coefficient $a_1 = 0$, which corresponds to $f_c(0, 0) = 0$. In this case, Eq. (3.13) simplifies to

$$\dot{x} = a_2x^2 + a_3xc + a_4c^2. \tag{3.16}$$

Furthermore, if $a_2 \neq 0$ and $a_3 \neq 0$, then at the nonhyperbolic equilibrium point $(x, c) = (0, 0)$, the system undergoes a bifurcation known as a *transcritical bifurcation*. Let us now analyze this bifurcation in detail.

Simple case: $a_4 = 0$

For now, let us assume that $a_4 = 0$. Then, Eq. (3.13) reduces to

$$\dot{x} = a_2 x^2 + a_3 x c. \tag{3.17}$$

The substitutions $\tilde{x} = |a_2| x$ and $\tilde{c} = |a_3| c$ transform this equation into:

$$\dot{x} = \frac{\dot{\tilde{x}}}{|a_2|} = a_2 \frac{\tilde{x}^2}{|a_2|^2} + a_3 \frac{\tilde{x}}{|a_2||a_3|} \tilde{c}.$$

Multiplying through by $|a_2|$, we get

$$\dot{\tilde{x}} = \frac{a_2}{|a_2|} \tilde{x}^2 + \frac{a_3}{|a_3|} \tilde{x} \tilde{c}.$$

Thus, we arrive at the equation

$$\dot{\tilde{x}} = \pm \tilde{x}^2 \pm \tilde{x} \tilde{c},$$

where the signs correspond to the signs of a_2 and a_3.

After reverting to the original notations (removing tildes), the equation becomes

$$\dot{x} = \pm x^2 \pm x c. \tag{3.18}$$

This equation is the normal form for a *transcritical bifurcation*. There are four possible cases depending on the signs of the terms in the right-hand side. Let us first consider the case when both terms are positive

$$\dot{x} = x^2 + x c.$$

To find the equilibrium points, we solve

$$x^2 + x c = 0. \tag{3.19}$$

The solutions of Eq. (3.19) are

$$X_1 = 0 \quad \text{and} \quad X_2 = -c.$$

For the stability analysis, we compute the derivative of $f(x, c) = x^2 + x c$ with respect to x

$$f_x(x,c) = 2x + c.$$

Evaluating this at $x = 0$ gives

$$f_x(0,c) = c,$$

which means that the equilibrium point $X_1 = 0$ is stable when $c < 0$ (since $f_x < 0$), and unstable when $c > 0$.

Similarly, evaluating at $x = -c$ gives

$$f_x(-c,c) = -2c + c = -c,$$

so the equilibrium point $X_2 = -c$ is stable when $c > 0$ and unstable when $c < 0$.

Thus, at $c = 0$, the two equilibria coincide at $x = 0$ and exchange their stability as c crosses zero. This is the characteristic behavior of a transcritical bifurcation.

The bifurcation diagrams for all four possible cases (corresponding to various sign combinations in the normal form Eq. (3.18)) are shown in Figure 3.9.

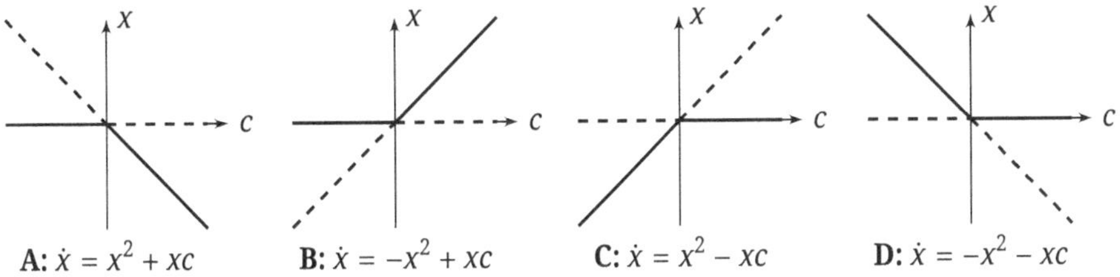

Figure 3.9: Bifurcation diagrams for the four cases of a transcritical bifurcation. Stable equilibria are shown with solid lines and unstable equilibria with dashed lines.

Case $a_4 \neq 0$

In this case, the equilibria of the flow (3.16) are determined by the roots of the quadratic equation

$$a_2x^2 + a_3xc + a_4c^2 = 0.$$

Solving this equation yields

$$X_{1,2} = \frac{-a_3c \pm \sqrt{(a_3c)^2 - 4a_2a_4c^2}}{2a_2} = \frac{-a_3 \pm \sqrt{a_3^2 - 4a_2a_4}}{2a_2}c.$$

If the discriminant $a_3^2 - 4a_2a_4$ is negative, then for $c = 0$ the only root is $x = 0$, and for all other values of c, there are no real roots. In this scenario, no (local) bifurcation occurs at $(0,0)$.

On the other hand, if the discriminant is positive ($a_3^2 - 4a_2a_4 > 0$), there are always two real roots that coincide at $x = 0$ when $c = 0$. We can express them as

$$X_{1,2} = \beta_{1,2}c, \quad \text{where } \beta_{1,2} = \frac{-a_3 \pm \sqrt{a_3^2 - 4a_2a_4}}{2a_2}.$$

Thus, the flow (3.16) can be factorized as

$$\dot{x} = a_2x^2 + a_3xc + a_4c^2 = a_2(x - \beta_1c)(x - \beta_2c).$$

Introducing the substitutions $\tilde{x} = |a_2|(x - \beta_1c)$ and $\tilde{c} = |\beta_1 - \beta_2|c$, we obtain

$$\dot{\tilde{x}} = \pm\tilde{x}^2 \pm \tilde{x}\tilde{c},$$

which is identical to the normal form Eq. (3.18) for the transcritical bifurcation.

Theorem 3.4. *Consider the flow $\dot{x} = f(x, c)$. If $f(x, c) = xg(x, c)$ and $g(0, 0) = 0$, then the flow has a transcritical bifurcation at the point $(x, c) = (0, 0)$.*

Proof. Indeed,

$$f(0, 0) = 0 * g(0, 0) = 0,$$

so we have an equilibrium at $(0, 0)$.

Furthermore,

$$f_x(x, c) = \frac{\partial}{\partial x}(xg(x, c)) = g(x, c) + xg_x(x, c),$$

which is again zero when $x = 0$ and $c = 0$. Therefore, $(0, 0)$ is a nonhyperbolic equilibrium.

Now we note that

$$\frac{\partial f(x, c)}{\partial c} = (xg(x, c))_c = xg_c(x, c) = 0$$

at (0, 0) and similarly,

$$\frac{\partial^2 f(x, c)}{\partial c^2} = (xg(x, c))_{cc} = xg_{cc}(x, y) = 0$$

at $(0, 0)$. Thus, $a_1 = 0$ and $a_4 = 0$, and we arrive at Eq. (3.17), which describes the flow with transcritical bifurcation at $(0, 0)$. □

Example 3.5. Analyze bifurcation taking place in the flow

$$\dot{x} = x(r - e^x), \quad x \in \mathbb{R}.$$

Solution:
Bifurcations take place at nonhyperbolic fixed points, that is, at points where $f(x, r) = 0$ and $f_x(x, r) = 0$. Let us start with the first condition

$$f(x, r) = 0 \rightarrow x(r - e^x) = 0,$$

from which it follows that there are two fixed points: $X_1 = 0$, which exists for all values of r, and $X_2 = \ln r$, which exists only for $r > 0$.

It is easy to see that at $r = 1$, these two fixed points coincide. Also, from

$$f_x(x, r) = r - xe^x - e^x$$

we have $f_x(x = 0, r) = r - 1 = 0$ when $r = 1$. Moreover, $X_1 = 0$ is stable if $r < 1$ and unstable otherwise. Thus, at $r = 1$, we have a *transcritical bifurcation.*

For completeness, one can analyze the case when $f_x(x = \ln r, r) = 0$, which leads to $r = 1$ or $r = 0$. The case $r = 1$ has already been analyzed, while $r = 0$ leads to $X_2 = -\infty$, which is a degenerate case.

3.5.3 Pitchfork bifurcation

We still consider the flow $\dot{x} = f(x, c)$, which has a nonhyperbolic equilibrium at $(0, 0)$, so that its second-order approximation is given by Eq. (3.13).

We already know the following:
- If $a_1 \neq 0$ and $a_2 \neq 0$, then at $(0, 0)$ we have a fold bifurcation.
- If $a_1 = 0$, $a_2 \neq 0$, $a_3 \neq 0$, and $a_3^2 - 4a_2a_4 > 0$, then at $(0, 0)$ we have a transcritical bifurcation.

Now, let us assume that $a_1 = 0$ and $a_2 = 0$. Then, Eq. (3.13) reduces to

$$\dot{x} = a_3xc + a_4c^2.$$

If $c = 0$, then every point is an equilibrium point, since $f(x, 0) = 0\ \forall x$.

This implies that the equilibrium at $(0, 0)$ is not isolated, and thus the behavior is not local. Therefore, we need to extend our approximation to the third order, which is given by

$$\dot{x} = a_3xc + a_4c^2 + a_5x^3 + a_6x^2c + a_7xc^2 + a_8c^3,$$

where

$$a_5 = \frac{f_{xxx}(0, 0)}{3!}, \quad a_6 = \frac{f_{xxc}(0, 0)}{2!}, \quad a_7 = \frac{f_{xcc}(0, 0)}{2!}, \quad a_8 = \frac{f_{ccc}(0, 0)}{3!}.$$

Since, in Taylor approximations, we consider $|x| \ll 1$ and $|c| \ll 1$, the higher-order terms satisfy the following inequalities:

$$|a_8c^3| \ll |a_4c^2|, \quad |a_7xc^2| \ll |a_3xc|, \quad |a_6x^2c| \ll |a_3xc|.$$

Thus, we can neglect these terms and approximate the flow by

$$\dot{x} = a_3xc + a_4c^2 + a_5x^3.$$

Third-order approximation makes sense if at least one of the third-order terms is of the same order as one of the second-order terms. This condition is satisfied if $c \sim x^2$, since under this assumption we have $xc \sim x^3$ and therefore $a_3xc \sim a_5x^3$.

At the same time, since $c^2 \sim x^4$, we obtain $|a_4c^2| \ll |a_5x^3|$. Therefore, we can neglect the a_4c^2 term, and our third-order approximation reduces to

$$\dot{x} = a_3xc + a_5x^3. \tag{3.20}$$

Let $\tilde{x} = \sqrt{|a_5|}x$ and $\tilde{c} = |a_3|c$. Then,

$$\frac{\dot{\tilde{x}}}{\sqrt{|a_5|}} = a_3\frac{\tilde{x}}{\sqrt{|a_5|}} \cdot \frac{\tilde{c}}{|a_3|} + a_5\frac{\tilde{x}^3}{(\sqrt{|a_5|})^3}$$

or, equivalently,

$$\dot{\tilde{x}} = \frac{a_3}{|a_3|}\tilde{x}\tilde{c} + \frac{a_5}{|a_5|}\tilde{x}^3.$$

Thus, after changing notations ($\tilde{x} \to x$ and $\tilde{c} \to c$), we have

$$\dot{x} = \pm xc \pm x^3, \tag{3.21}$$

which is the normal form for the *pitchfork bifurcation*.

Similarly to the fold and transcritical bifurcations, we have four different cases to consider, determined by the combinations of signs in Eq. (3.21).

Case 1. Consider the equation $\dot{x} = xc + x^3$.

The equilibrium points are the roots of

$$0 = x(c + x^2) \quad \Rightarrow \quad X_1 = 0, \quad X_{2,3} = \pm\sqrt{-c}.$$

The solution X_1 exists always, while the solutions $X_{2,3}$ exist only if $c < 0$.
To analyse the stability of equilibria we need the derivative $f_x(x, c) = c + 3x^2$.
For the equilibrium point $X_1 = 0$, we have:

$$f_x\,(x = 0, c) = c.$$

Therefore, $X_1 = 0$ is *stable* if $c < 0$ and *unstable* if $c > 0$.

The derivative $f_x(x,c)$ at $X_2 = \sqrt{-c}$:

$$f_x(x = \sqrt{-c}, c) = c + 3x^2 = c + 3(-c) = -2c.$$

Since $c < 0$, we have $-2c > 0$, which implies that $X_2 = \sqrt{-c}$ is *unstable.*
Similarly, evaluate at $X_3 = -\sqrt{-c}$

$$f_x(x = -\sqrt{-c}, c) = c + 3x^2 = c + 3(-c) = -2c > 0.$$

Thus, $X_3 = -\sqrt{-c}$ is also *unstable*.

Case 2. Now consider $\dot{x} = -xc + x^3$ and find the equilibria

$$0 = -xc + x^3 \quad \Longrightarrow \quad X_1 = 0, \quad X_{2,3} = \pm\sqrt{c}.$$

Now, $X_{2,3}$ exist if and only if $c > 0$.
We evaluate $f_x = -c + 3x^2$ at all equilibrium points as follows:
At $X_1 = 0$,

$$f_x(x = 0, c) = -c < 0,$$

hence, $X_1 = 0$ is stable when $c > 0$ and unstable otherwise.
At $X_2 = \sqrt{c}$,

$$f_x(x = \sqrt{c}, c) = -c + 3c = 2c > 0, \quad \text{as } c > 0,$$

thus, $X_2 = \sqrt{c}$ is unstable.
At $X_3 = -\sqrt{c}$,

$$f_x(x = -\sqrt{c}, c) = -c + 3c = 2c > 0, \quad \text{as } c > 0,$$

thus, $X_3 = -\sqrt{c}$ is also unstable.

Case 3. Now consider $\dot{x} = xc - x^3$ and find the equilibria

$$0 = xc - x^3 \quad \Longrightarrow \quad X_1 = 0, \quad X_{2,3} = \pm\sqrt{c}.$$

$X_{2,3}$ exist if and only if $c > 0$.
We evaluate $f_x = c - 3x^2$ at all equilibrium points as follows:
At $X_1 = 0$,

$$f_x(x = 0, c) = c > 0,$$

hence, $X_1 = 0$ is unstable when $c > 0$ and stable otherwise.

At $X_2 = \sqrt{c}$,

$$f_x(x = \sqrt{c}, c) = c - 3c = -2c < 0, \quad \text{as } c > 0,$$

thus, $X_2 = \sqrt{c}$ is stable.

At $X_3 = -\sqrt{c}$,

$$f_x(x = -\sqrt{c}, c) = c - 3c = -2c < 0, \quad \text{as } c > 0,$$

thus, $X_3 = -\sqrt{c}$ is also stable.

Case 4. Now, consider $\dot{x} = -xc - x^3$. The equilibria are found as

$$0 = -xc - x^3 \quad \Longrightarrow \quad X_1 = 0, \quad X_{2,3} = \pm\sqrt{-c}.$$

$X_{2,3}$ exist if and only if $c < 0$.

We evaluate $f_x = -c - 3x^2$ at all equilibrium points as follows:

At $X_1 = 0$,

$$f_x(x = 0, c) = -c > 0, \quad \text{as } c < 0,$$

hence, $X_1 = 0$ is unstable if $c < 0$ and stable otherwise.

At $X_2 = \sqrt{-c}$,

$$f_x(x = \sqrt{-c}, c) = -c - 3(-c) = 2(-c) < 0, \quad \text{as } c < 0,$$

thus, $X_2 = \sqrt{-c}$ is stable.

At $X_3 = -\sqrt{-c}$,

$$f_x(x = -\sqrt{-c}, c) = -c - 3(-c) = 2(-c) < 0, \quad \text{as } c < 0,$$

thus, $X_3 = -\sqrt{-c}$ is also stable.

Bifurcation diagrams for all four cases are shown in Figure 3.10. In Cases 1 and 2, equilibria $X_{2,3}$ are both unstable, and these cases are referred to as *subcritical pitchfork bifurcation*. In Cases 3 and 4, equilibria $X_{2,3}$ are both stable, and these cases are referred to as *supercritical pitchfork bifurcation*.

Example 3.6. Analyze bifurcation taking place in the flow

$$\dot{x} = x + \frac{rx}{1 + x^2}, \quad x, r \in \mathbb{R}.$$

Solution:

Fixed points:

$$f(x, r) = 0 \quad \Rightarrow \quad x + \frac{rx}{1 + x^2} = 0 \quad \Rightarrow \quad X_1 = 0,$$

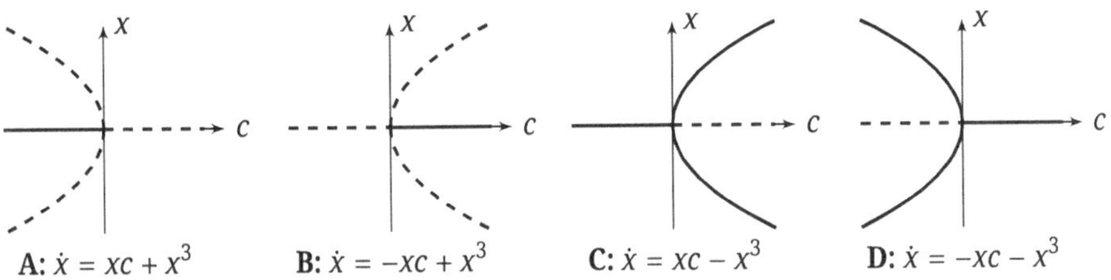

Figure 3.10: Bifurcation diagrams for the four possible cases of a pitchfork bifurcation. Panels **A** and **B** illustrate the *subcritical pitchfork* bifurcation, in which a stable equilibrium is flanked by two unstable equilibria. Panels **C** and **D** show the *supercritical pitchfork* bifurcation, where an unstable equilibrium is surrounded by two stable equilibria.

which exists for all values of r, and

$$X_{2,3} = \pm\sqrt{-(1+r)},$$

which exist only when $r \leq -1$.

At $r = -1$, all three fixed points coincide, $X_1 = X_2 = X_3 = 0$, which clearly indicates a pitchfork bifurcation.

Furthermore,

$$f_x'(x,r) = 1 + \frac{r}{1+x^2} + \frac{2rx^2}{(1+x^2)^2},$$

and

$$f_x'(0,r) = 1 + r < 0 \quad \text{if} \quad r < -1.$$

From this, it follows that $X_{2,3}$ are stable, and we therefore have a *supercritical pitchfork bifurcation.*

Theorem 3.5. *Consider the flow $\dot{x} = f(x,c)$. If $f(x,c)$ is an odd function with respect to x and $f_x(0,0) = 0$, then the flow has a pitchfork bifurcation at the point $(x,c) = (0,0)$.*

Proof. Let us consider the third-order Maclaurin expansion for the flow

$$\begin{aligned}\dot{x} = f(x,c) = f(0,0) + f_x(0,0)\,x + f_c(0,0)\,c + \frac{1}{2}f_{xx}(0,0)\,x^2 \\ + f_{cx}(0,0)\,xc + \frac{1}{2}f_{cc}(0,0)\,c^2 + \frac{1}{6}f_{xxx}(0,0)\,x^3 \\ + \frac{1}{2}f_{xxc}(0,0)\,x^2c + \frac{1}{2}f_{xcc}(0,0)\,xc^2 + \frac{1}{6}f_{ccc}(0,0)\,c^3.\end{aligned}$$

Since $f(x,c)$ is an odd function with respect to x, that is, $f(-x,c) = -f(x,c)$ for all x, the following properties hold:

- Since $f(x,c)$ is odd in x, then $f(0,c) = 0$ for all c. In particular, $f(0,0) = 0$. Combined with the condition $f_x(0,0) = 0$, this implies that at $(x,c) = (0,0)$ the system has a nonhyperbolic equilibrium.
- The derivative $f_c(x,c)$ is also odd in x

$$\begin{aligned} f_c(x,c) &= \frac{f(x,c+\delta c) - f(x,c)}{\delta c} = \frac{-f(-x,c+\delta c) + f(-x,c)}{\delta c} \\ &= \frac{-(f(-x,c+\delta c) - f(-x,c))}{\delta c} = -f_c(-x,c). \end{aligned}$$

  This implies $f_c(0,c) = 0$ for all c, and in particular $f_c(0,0) = 0$.
- Since $f_c(x,c)$ is odd in x, it follows similarly that $f_{cc}(x,c)$ and $f_{ccc}(x,c)$ are also odd in x. Therefore, $f_{cc}(0,0) = 0$, and $f_{ccc}(0,0) = 0$.
- The derivative $f_x(x,c)$ is even in x

$$\begin{aligned} f_x(x,c) &= \frac{f(x+\delta x,c) - f(x,c)}{\delta x} = \frac{-f(-x-\delta x,c) + f(-x,c)}{\delta x} \\ &= \frac{f(-x,c) - f(-x-\delta x,c))}{\delta x} = f_x(-x,c). \end{aligned}$$

  So, $f_x(-x,c) = f_x(x,c)$.
- The second derivative $f_{xx}(x,c)$ is odd in x

$$\begin{aligned} f_{xx}(x,c) &= \frac{f_x(x+\delta x,c) - f_x(x,c)}{\delta x} = \frac{f_x(-x-\delta x,c) + f_x(-x,c)}{\delta x} \\ &= -\frac{f_x(-x,c) - f_x(-x-\delta x,c))}{\delta x} = -f_{xx}(-x,c), \end{aligned}$$

  so $f_{xx}(0,c) = 0$, and in particular $f_{xx}(0,0) = 0$.
- As a consequence of $f_{xx}(x,c)$ being odd in x, we get $f_{xxc}(x,c)$ is odd in x, and thus $f_{xxc}(0,c) = 0$.

Thus, the third-order Maclaurin expansion of $\dot{x} = f(x,c)$ is reduced to

$$\dot{x} = f_{cx}xc + f_{xxx}(0,0)\frac{x^3}{6} + f_{xcc}(0,0)\frac{xc^2}{2}$$

as all other terms are equal to zero. Since $c^2 \ll |c|$, the last term is negligible compared to the first, and thus

$$\dot{x} = x\left(f_{cx}(0,0)c + f_{xxx}(0,0)\frac{x^2}{6}\right)$$

which is in the form of the pitchfork bifurcation Eq. (3.20), and hence we conclude that a pitchfork bifurcation occurs at the point $(0,0)$. □

3.6 Multistability

We have seen that, in the course of a pitchfork bifurcation, two stable fixed points appear. Each of these fixed points has its own domain of attraction, and this means that the long-time behavior observed in such a system strongly depends on the initial conditions. Orbits starting in different domains of attraction converge to different fixed points. This phenomenon, known as *multistability*, is very important in many practical settings. In biology, it is often associated with mechanisms that allow a switch from one steady state to another, known as *biological switches*. The simplest flow allowing two steady states is generated by a cubic polynomial as evident from the fact that only a cubic approximation of a map with a parameter can exhibit a pitchfork bifurcation. However, multistability is not necessarily associated with a supercritical pitchfork bifurcation.

Example 3.7. For the flow

$$\dot{x} = x(x-1)(2-x) - c, \tag{3.22}$$

defined for $x, c \in \mathbb{R}$, find all bifurcation points and determine the range of parameter values for which the system is multistable.

Solution:
The first step is to find the nonhyperbolic equilibria, that is, the solutions of the system $f(x,c) = 0$ and $f_x(x,c) = 0$.

$$f(x,c) = (x^2 - x)(2-x) - c = -x^3 + 3x^2 - 2x - c \quad \Rightarrow \quad f_x(x,c) = -3x^2 + 6x - 2.$$

Let $f_x(x,c) = 0$:

$$-3x^2 + 6x - 2 = 0.$$

Roots of this equation are

$$X_{1,2} = \frac{6 \pm \sqrt{36 - 4(3)(2)}}{6} = \frac{6 \pm \sqrt{12}}{6} = \frac{3 \pm \sqrt{3}}{3}.$$

To find the non-hyperbolic equilibria we substitute these roots into $f(x,c) = 0$. Substituting $x = X_1 = \frac{3+\sqrt{3}}{3}$, we obtain

$$\left(\frac{3+\sqrt{3}}{3}\right)\left(\frac{3+\sqrt{3}}{3} - 1\right)\left(2 - \frac{3+\sqrt{3}}{3}\right) - C_1 = 0 \quad \Rightarrow \quad C_1 = \frac{2\sqrt{3}}{9}$$

So, the first nonhyperbolic equilibrium point is

$$(X_1, C_1) = \left(\frac{3+\sqrt{3}}{3}, \frac{2\sqrt{3}}{9}\right).$$

Similarly, for $X_2 = \frac{3-\sqrt{3}}{3}$, we find

$$(X_2, C_2) = \left(\frac{3-\sqrt{3}}{3}, -\frac{2\sqrt{3}}{9}\right).$$

The second-order approximation of the flow near the bifurcation point (X_1, C_1) is

$$\dot{x} = f_c(c - C_1) + f_{xx}\frac{(x - X_1)^2}{2} + f_{xc}(x - X_1)^2(c - C_1)^2 + f_{cc}\frac{(c - C_1)^2}{2}.$$

Now, given the function $f(x, c) = x(x-1)(2-x) - c$, we compute the partial derivatives

$$\begin{aligned} f_c(x, c) &= -1, \\ f_x(x, c) &= -3x^2 + 6x - 2, \\ f_{xx}(x, c) &= -6x + 6, \\ f_{xc}(x, c) &= 0, \\ f_{cc}(x, c) &= 0. \end{aligned}$$

At the bifurcation point $(X_1, C_1) = (\frac{3+\sqrt{3}}{3}, \frac{2\sqrt{3}}{9})$, we already know that $f_x(X_1, C_1) = 0$ (non-hyperbolic equilibrium).

Now compute $f_{xx}(X_1, C_1)$:

$$f_{xx}(C_1, X_1) = -6X_1 + 6 = -2\sqrt{3}.$$

Therefore, the second-order approximation simplifies to

$$\dot{x} = -(c - C_1) - 2\sqrt{3}\frac{(x - X_1)^2}{2} = -(c - C_1) - \sqrt{3}(x - X_1)^2.$$

Let $\tilde{x} = x - X_1$ and $\tilde{c} = c - C_1$. Then the second-order approximation becomes

$$\dot{x} = \dot{\tilde{x}} = -\tilde{c} - \sqrt{3}\tilde{x}^2.$$

Now, introduce the substitutions $y = \sqrt{3}\tilde{x}$ and $r = \sqrt{3}\tilde{c}$, to yield

$$\dot{y} = -r - y^2.$$

This is the normal form of a fold bifurcation. The bifurcation diagram corresponding to this case is shown in Figure 3.8D.

Repeating the procedure, we find that

$$f_{xx}(C_2, X_2) = -6X_2 + 6 = 2\sqrt{3},$$

which gives the second-order approximation of the flow at the bifurcation point (X_2, C_2) as

$$\dot{x} = -(c - C_2) + \sqrt{3}(x - X_2)^2.$$

Introducing the substitutions $y = \sqrt{3}(x - X_2)$ and $r = \sqrt{3}(c - C_2)$, we obtain

$$\dot{y} = -r + y^2.$$

This again corresponds to a fold bifurcation, but with reversed stability structure compared to the previous case. The bifurcation diagram for this case is shown in Figure 3.8B.

The complete bifurcation diagram, combining both bifurcations occurring in the system, is presented in Figure 3.11. As is evident from the diagram, the system is multistable when $c \in (C_2, C_1)$. In this range of parameter values, there are two stable fixed points separated by an unstable steady state.

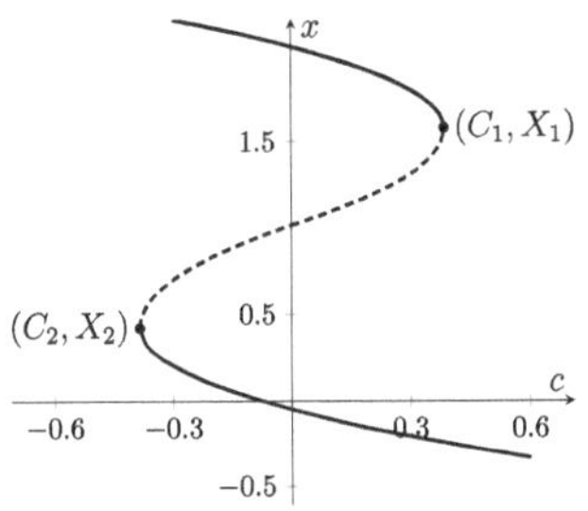

Figure 3.11: Bifurcation diagram for the dynamical system in Example 3.7. Two bifurcation points are present, each corresponding to a fold bifurcation. The normal form at the point (X_1, C_1) is $\dot{y} = -r - y^2$, while at the point (X_2, C_2) it is $\dot{y} = -r + y^2$.

The dynamical regime exhibited by a system with a parameter is often controlled by varying the value of that parameter. Assume that $c \in (C_2, C_1)$ and the system is in the steady state on the upper branch $x = x_1(c)$. A gradual increase in c will then lead to a sharp switch from the state $x = X_1$ to a lower branch $x = x_2(c)$ when c reaches the value $c = C_1$. Similarly, if the system is in the steady state on the lower branch $x = x_2(c)$, a gradual decrease in c beyond the interval (C_2, C_1) will result in a corresponding switch from $x = X_2$ to the upper branch $x = x_1(c)$ when c reaches $c = C_2$. Thus, if $c \in (C_2, C_1)$, the state of the system switches whenever the parameter c crosses either the left or the right boundary of this interval. Such a switch is an important mechanism of functional regulation in biological systems.

Also, if c is outside the interval (C_2, C_1) and enters it from above, then $x = x_2(c)$ for all $c \in (C_2, C_1)$. However, if c enters this interval from below, then $x = x_1(c)$ for all $c \in (C_2, C_1)$. This phenomenon is known in physics as *hysteresis*.

Thus, multistability is a phenomenon associated with the coexistence of several attractors and, as such, can be observed only in nonlinear systems. As we have seen, multistability can lead to abrupt transitions between qualitatively different states of the system when parameters or initial conditions are varied. In the following chapters we will consider this phenomenon further in multidimensional systems.

3.7 Summary

In this chapter, we introduced the one-dimensional autonomous flow $\dot{x} = f(x)$ and a qualitative method for its analysis. In particular, we showed that

- The flow admits stationary solutions, which are given by the roots of the algebraic equation $f(x) = 0$.
- The stationary solution $x(t) = X$ is stable if $f'(X) < 0$ and unstable if $f'(X) > 0$.
- All solutions are monotonic, with their direction determined by the sign of $f(x)$ along the trajectory.

We have also considered flows with a parameter, $\dot{x} = f(x, c)$ and introduced bifurcation values of the parameter c, which are associated with changes in the number and/or stability of stationary solutions. We have shown that a local bifurcation can occur at the point (X, C) if $f(X, C) = 0$ (equilibrium point) and $f_x(X, C) = 0$, (the equilibrium point is non-hyperbolic). We have identified three types of local bifurcations that can take place in 1D flows as follows:

- *Fold bifurcation*—occurs if $f_c(X, C) \neq 0$ and $f_{xx}(X, C) \neq 0$. At the bifurcation point, two equilibrium points (one stable and one unstable) appear or disappear.
- *Transcritical bifurcation*—occurs if $f_c(X, C) = 0$, $f_{xx}(X, C) \neq 0$ and $f_{xc}(X, C) \neq 0$. Two equilibria (one stable, one unstable) collide and exchange stability.
- *Pitchfork bifurcation*—occurs if $f_c(X, C) = 0$, $f_{xx}(X, C) = 0$, $f_{xc}(X, C) \neq 0$, and $f_{xxx}(X, C) \neq 0$. An equilibrium point changes stability, while two additional equilibria appear or disappear at its location. In the *supercritical* case, the new equilibria are both stable; in the *subcritical* case, they are both unstable, with stability opposite to that of the original equilibrium.
- If $f_c(X, C) = 0$, $f_{xx}(X, C) = 0$, $f_{xc}(X, C) = 0$, that is, none of these conditions for bifurcations arising in the second- or third-order approximations are satisfied, a local bifurcation can still occur at (X, C), determined from a higher-order approximation. However, it will still belong to one of the three types described.

As we remember, bifurcations in maps take place when the derivative of the map-generating function at the equilibrium point is either 1 or −1, while here we have shown that for flows this derivative should be equal to 0. To explain this discrepancy between maps and flows, we note that the derivative is defined as

$$\dot{x} = \lim_{\Delta t \to 0} \frac{\Delta x}{\Delta t}.$$

If we discretize time by taking a small step Δt, then the flow $\dot{x} = f(x, c)$ generates a map

$$x_{n+1} = x_n + \Delta x = x_n + \dot{x}\,\Delta t = x_n + f(x_n, c)\,\Delta t.$$

Thus, the flow $\dot{x} = f(x, c)$ is associated with a map $x_{n+1} = g(x_n, c)$ with a map-generating function $g(x, c) = x + f(x, c)\Delta t$. When in the flow $f'_x(x, c) = 0$, in the corresponding map, we obtain

$$g'_x(x, c) = 1 + f'_x(x, c)\Delta t = 1.$$

Thus, we have essentially the same condition for the occurrence of fold, transcritical or pitchfork bifurcations in the flow and in the associated map.

This is not true for a flip bifurcation, which takes place in a map when $g'(x) = -1$. This implies that

$$1 + f'(x)\Delta t = -1, \quad \text{or} \quad \Delta t = -\frac{2}{f'(x)}.$$

This means that a flip bifurcation occurs in the map associated with a flow only for a finite value of the time step Δt, which takes us beyond the differential-equation limit. This explains why we do not have flip bifurcations in flows.

Additional reading on the theory of flows, and especially on bifurcation theory, is provided in the textbooks [14, 15]. Further problems for practicing the quantitative analysis techniques demonstrated in this chapter can be found in [8].

3.8 Exercises

Problem 3.1. Determine the equilibrium points and their stability for the following differential equation:

$$\dot{n} = -n^2 + 15n - 50.$$

Integrate this equation (using partial fractions) to find $n(t)$, assuming $n(0) = 7$. What happens to the value of n as $t \to \infty$?

Problem 3.2. Consider the logistic equation with a parameter

$$\dot{y} = 3y(y - 10) - c.$$

What kind of bifurcation takes place in this equation? Plot the bifurcation diagram and obtain the normal form for the bifurcation.

Problem 3.3. Analyze bifurcations taking place in systems with parameter r described by the following equations:

(a) $\dot{x} = 1 + rx + x^2$;

(b) $\dot{x} = rx - \ln(1 + x)$, $x > -1$;

(c) $\dot{x} = rx - \sin x$, $-\pi < x < \pi$;

(d) $\dot{x} = r - \cosh x$;

(e) $\dot{x} = x - rx(1 - x)$;
(f) $\dot{x} = x - r \tan x$, $-\pi/2 < x < \pi/2$.

Problem 3.4. For the differential equations
(a) $\dot{x} = x^3 - 2x^2 + rx$;
(b) $\dot{x} = rx + x^3 - x^5$,

defined for $x, r \in \mathbb{R}$,
(i) Find all the stationary solutions and analyse their stability.
(ii) Use appropriate approximations to determine the normal forms of all bifurcations taking place in these systems.
(iii) Draw the bifurcation diagrams.

3.9 Solutions to exercises

Solution to Problem 3.1. Equilibrium points are the roots of the equation $-n^2 + 15n - 50 = 0$, which are $N_1 = 5$ and $N_2 = 10$.

To analyse the stability of these equilibrium states, we find the derivative $f'(n) = -2n + 15$ at the equilibrium values of n.

$f'(N_1) = 5 > 0$, i. e. $N_1 = 5$ is unstable, and $f'(N_2) = -5 < 0$, thus $N_2 = 10$ is stable. The basin of attraction of the stable equilibrium is the interval $(5, \infty)$.

To integrate the equation, we separate variables

$$\frac{dn}{(n-5)(10-n)} = dt$$

To integrate the left-hand side, we use the method of partial fractions

$$\frac{1}{(n-5)(10-n)} = \frac{1}{5}\left(\frac{1}{n-5} + \frac{1}{10-n}\right).$$

Thus,

$$\frac{1}{5}(\ln(n-5) - \ln(10-n)) = t + c,$$

or equivalently,

$$\ln\left(\frac{n-5}{10-n}\right) = 5t + c_1.$$

Finally,

$$\frac{n-5}{10-n} = A\exp(5t).$$

For the initial condition $n(0) = 7$,

$$A = \frac{n(0) - 5}{10 - n(0)} = \frac{7 - 5}{10 - 7} = \frac{2}{3}.$$

The explicit solution is

$$n(t) = \frac{10 + 7.5\exp(-5t)}{1 + 1.5\exp(-5t)}.$$

Note that

$$\lim_{t\to\infty} n(t) = 10.$$

Solution to Problem 3.2. The equilibrium points

$$Y_{1,2} = 5 \pm \sqrt{25 + \frac{c}{3}}$$

exist when $c > -75$. Thus, depending on the value of the parameter c, there can be either two or no equilibrium points—a clear indication of a fold bifurcation. The roots coincide and are equal to 5 when $c = -75$. Thus, $(C, Y) = (-75, 5)$ is a bifurcation point.

To determine the stability of the equilibrium points, we compute the derivative of the right-hand side of the given equation

$$f_y'(y) = 6y - 30$$

and obtain, for $c > -75$,

$$f_y'\left(Y_1 = 5 + \sqrt{25 + \frac{c}{3}}\right) = 6\left(5 + \sqrt{25 + \frac{c}{3}}\right) - 30 = 6\sqrt{25 + \frac{c}{3}} > 0.$$

Thus, Y_1 is unstable (and correspondingly Y_2 is stable). The bifurcation diagram is shown in Figure 3.12.

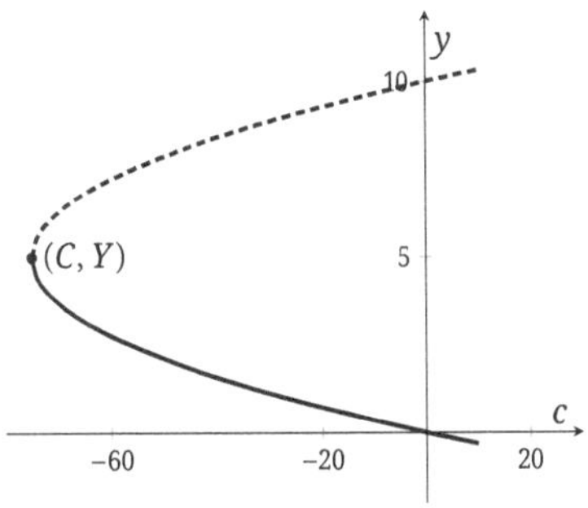

Figure 3.12: Bifurcation diagram for the flow $\dot{y} = 3y(y - 10) - c$. The bifurcation point $(C, Y) = (-75, 5)$ (marked by a dot) corresponds to a fold bifurcation.

To shift the bifurcation point $(5, -75)$ to the origin $(0, 0)$, make the following substitutions: $\tilde{y} = y - 5$ and $\tilde{c} = c + 75$, which leads to

$$\frac{d\tilde{y}}{dt} = 3\tilde{y}^2 - \tilde{c}.$$

Make a further change of variables, $x = 3\tilde{y}$ and $b = 3\tilde{c}$, to obtain the normal form

$$\frac{dx}{dt} = x^2 - b.$$

Solution to Problem 3.3.

(a) $\dot{x} = 1 + rx + x^2$, equilibrium points are the roots of $1 + rx + x^2 = 0$

$$X_{1,2} = \frac{-r \pm \sqrt{r^2 - 4}}{2},$$

which exist only for $|r| > 2$. At $r = \pm 2$, we have *fold bifurcations*.
For analysis of stability of equilibria we compute the derivative:

$$f(x, r) = 1 + rx + x^2 \quad \Rightarrow \quad f_x(x, r) = 2x + r$$

For $X_1 = \frac{-r+\sqrt{r^2-4}}{2}$, $f_x(X_1, r) = \sqrt{r^2 - 4} > 0$, hence X_1 is unstable.
For $X_2 = \frac{-r-\sqrt{r^2-4}}{2}$, $f_x(X_2, r) = -\sqrt{r^2 - 4} < 0$, hence X_2 is stable.

(b) $\dot{x} = rx - \ln(1+x)$, equilibrium points are the roots of $rx - \ln(1+x) = 0$. Graphically, it can be seen that rx and $\ln(1 + x)$ can have no more than two points of intersection. Taking the second-order approximation of $\ln(1 + x)$, we obtain:

$$f(x) = rx - \ln(1 + x) \approx rx - x + \frac{x^2}{2}.$$

Roots of $f(x) = 0$,

$$X_1 = 0, \quad X_2 = 2(1 - r).$$

At $r = 1$, we have $X_1 = X_2$—this is a *transcritical bifurcation*.
For analysis of stability of equilibria we compute the derivative:

$$f(x, r) = rx - x + \frac{x^2}{2} \quad \Rightarrow \quad f_x(x, r) = r - 1 + x.$$

For $X_1 = 0$: $f_x(X_1, r) = r - 1$—negative if $r < 1$, thus X_1 is stable if $r < 1$.
For $X_2 = 2(1 - r)$: $f_x(X_2, r) = 1 - r$—negative if $r > 1$, thus X_2 is stable when $r > 1$.

(c) $\dot{x} = rx - \sin x$, equilibrium points are the roots of $rx - \sin x = 0$. Graphically, it can be checked that rx and $\sin x$ can have no more than three points of intersection in the domain $(-\pi, \pi)$.

Taking the third-order approximation of $\sin x$,

$$f(x) = rx - \sin x \approx rx - x + \frac{x^3}{3!}.$$

The roots of $f(x, r) = 0$,

$$X_1 = 0; X_{2,3} = \pm\sqrt{6(1-r)},$$

with $X_{2,3}$ existing only when $r < 1$.
At $r = 1$, we have $X_1 = X_2 = X_3 = 0$—this corresponds to a pitchfork bifurcation.
For analysis of stability of equilibria we compute the derivative:

$$f(x,r) = rx - x + \frac{x^3}{3!} \quad \Rightarrow \quad f_x(x,r) = r - 1 + x^2/2.$$

For $X_1 = 0, f_{x,r}(0) = r - 1$, which is negative if $r < 1$, thus X_1 is stable when $r < 1$. Consequently, $X_{2,3}$ should be unstable (this can be verified directly), and we therefore have a *subcritical pitchfork bifurcation* at $r = 1$.

(d) Here, $\dot{x} = r - \cosh x$. Nonhyperbolic equilibrium points satisfy

$$f(r,x) = r - \cosh x = 0, \quad f_x(r,x) = \sinh x = 0.$$

The second condition gives $x = 0$, and substituting into $f(0, r) = 0$ yields $r = 0$. Thus the only non-hyperbolic equilibrium is at $(x, r) = (0, 0)$.
A second-order Taylor expansion of $f(x, r)$ at this point gives

$$\dot{x} = r - \frac{x^2}{2},$$

which is the normal form of a fold bifurcation. Hence, the system undergoes a *fold bifurcation* at $r = 0$.
The equilibria of the reduced equation are

$$X_{1,2} = \pm\sqrt{-2r},$$

which exist only for $r < 0$.
For $X_1 = -\sqrt{-2r}$, we have $f_x(X_1, r) = -X_1 > 0$, so X_1 is unstable.
For $X_2 = \sqrt{-2r}$, we have $f_x(X_2, r) = -X_2 < 0$, so X_2 is stable.

(e) Here, $\dot{x} = f(x, r)$ with $f(x, r) = x - rx(1 - x)$. Nonhyperbolic equilibria satisfy

$$f(r,x) = 0, \quad f_x'(r,x) = 0,$$

The equilibrium condition gives

$$x - rx(1-x) = x\big(1 - r(1-x)\big) = 0.$$

Thus, equilibria are

$$X_1 = 0, \quad 1 - r(1 - x) = 0 \quad \Rightarrow \quad X_2 = 1 - \frac{1}{r}.$$

A nonhyperbolic equilibrium condition,

$$f_x'(r, x) = 1 - r + 2rx = 0.$$

At $X_1 = 0$,

$$f_x'(r, 0) = 1 - r = 0 \quad \Rightarrow \quad r = 1.$$

At $X_2 = 1 - \frac{1}{r}$:

$$f_x\left(1 - \frac{1}{r}, r\right) = r - 1 = 0 \quad \Rightarrow \quad r = 1,$$

for which $X_2 = 0$. Thus the only non-hyperbolic equilibrium is at $(x, r) = (0, 1)$.
Also, at the equilibrium $X_1 = 0$:

$$f_x(0, r) = 1 - r < 0 \quad \Rightarrow \quad r > 1,$$

hence, it is stable if $r > 1$ and unstable if $r < 1$.
Similarly, at the equilibrium $X_2 = 1 - \frac{1}{r}$:

$$f_x\left(1 - \frac{1}{r}, r\right) = r - 1 < 0 \quad \Rightarrow \quad r < 1,$$

hence, it is stable if $r < 1$ and unstable if $r > 1$.
At the equilibrium $X_2 = \tilde{r}$,

$$f_x'(\tilde{r}, \tilde{r}) = 2\tilde{r} - \tilde{r} = \tilde{r}.$$

Hence, it is stable if $\tilde{r} < 0$ (i. e., $r < 1$) and unstable if $\tilde{r} > 0$ (i. e., $r > 1$).
Thus, the system undergoes a *transcritical bifurcation* at $(x, r) = (0, 1)$, with the two equilibrium branches exchanging stability.

(f) Nonhyperbolic equilibrium points satisfy

$$f(x, r) = x - r \tan x = 0, \quad f_x(x, r) = 1 - r \sec^2 x = 0.$$

The second condition gives

$$1 - r \sec^2 x = 0 \quad \Rightarrow \quad r = \cos^2 x.$$

Together with $x - r \tan x = 0$, the only solution is

$$x = 0, \quad r = 1.$$

To analyze the bifurcation, expand $\tan x$ to third order

$$\tan x = x + \frac{x^3}{3} + O(x^5).$$

Thus, the system becomes

$$\dot{x} = x - r\left(x + \frac{x^3}{3}\right) = (1 - r)x - \frac{r}{3}x^3.$$

The equilibria

$$X_1 = 0, \quad X_{2,3} = \pm\sqrt{3\left(1 - \frac{1}{r}\right)}.$$

The nonzero equilibria exist only for $r > 1$. At $r = 1$, all three equilibria collide at $x = 0$, indicating a pitchfork bifurcation

Next, compute the derivative

$$f_x(x, r) = 1 - r + rx^2.$$

At the equilibrium $X_1 = 0$,

$$f_x(0, r) = 1 - r.$$

Thus, $f_x(0, r) < 0$ if $r > 1$, so X_1 is stable for $r > 1$ and unstable for $r < 1$.

At the equilibria $X_{2,3}$,

$$f_x(X_{2,3}, r) = 1 - r + rX_{2,3}^2 = 1 - r + r \cdot 3\left(1 - \frac{1}{r}\right) = 2(r - 1) > 0 \quad (r > 1).$$

Thus, both $X_{2,3}$ are unstable for $r > 1$.

Therefore, the system undergoes a *subcritical pitchfork bifurcation* at $(x, r) = (0, 1)$.

Solution to Problem 3.4.

(a) $\dot{x} = x^3 - 2x^2 + rx$

(i) Stationary solutions are roots of

$$x^3 - 2x^2 + rx = 0,$$

that is, $X_1 = 0$, which exists for all values of r and two other solutions given by the roots of

$$x^2 - 2x + r = 0,$$

namely,

$$X_{2,3} = 1 \pm \sqrt{1-r},$$

which exist when $r \leq 1$.
Stability is defined by the sign of the derivative

$$f_x(x,r) = 3x^2 - 4x + r.$$

For $X_1 = 0$,

$$f_x(0,r) = r,$$

which is negative (X_1 is stable) for $r < 0$, and positive (X_1 is unstable) or $r > 0$.
For $X_{2,3} = 1 \pm \sqrt{1-r}$,

$$f_x(1 \pm \sqrt{1-r}, r) = 3(1 \pm \sqrt{1-r})^2 - 4(1 \pm \sqrt{1-r}) + r = 2 - 2r \pm 2\sqrt{1-r}.$$

It follows that $X_2 = 1 + \sqrt{1-r}$ is always unstable, while $X_3 = 1 - \sqrt{1-r}$ is stable for $0 < r < 1$ and unstable for $r < 0$.

(ii) Bifurcations take place at nonhyperbolic equilibriums ($f_x(x,r) = 0$). For $X_1 = 0$, we have

$$f_x(0,r) = r = 0$$

when $r = 0$, and so $(0,0)$ is a bifurcation point.
Near $(x,r) = (0,0)$, the second order approximation is

$$\dot{x} \approx rx - 2x^2.$$

With the substitution $y = 2x$. this becomes

$$\dot{y} = ry - y^2,$$

which is a normal form of a *transcritical bifurcation*.
For $X_{2,3} = 1 \pm \sqrt{1-r}$, nonhyperbolicity requires

$$f_x(1 \pm \sqrt{1-r}, r) = 2 - 2r \pm 2\sqrt{1-r} = 0,$$

which implies $(1-r)^2 = 1 - r$. The solutions are $r = 0$ and $r = 1$. Thus, we find the second bifurcation point $(1,1)$.
Near $(x,r) = (1,1)$, where $X_{2,3}$ appear/disappear, we make the substitutions $y = x - 1$ and $c = r - 1$. This yields

$$\dot{y} \approx y^2 + cy + c.$$

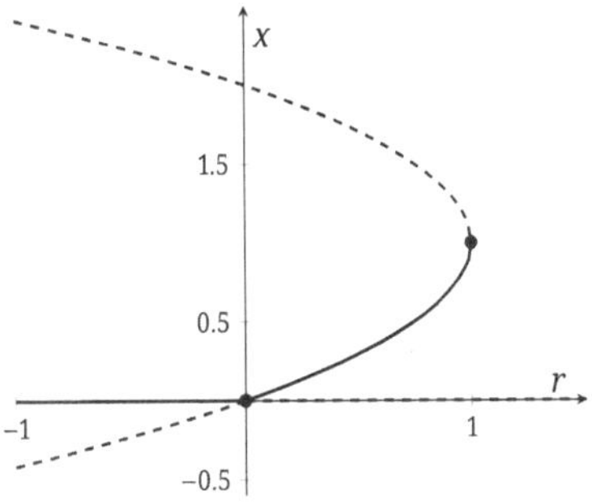

Figure 3.13: Bifurcation diagram for the flow $\dot{x} = x^3 - 2x^2 + rx$. Bifurcation points are marked by dots. The point $(r, x) = (1, 1)$ corresponds to a fold bifurcation, while the point $(r, x) = (0, 0)$ corresponds to a transcritical bifurcation.

Completing the square,

$$y^2 + cy + c = (y + c/2)^2 + (c - c^2/4).$$

With the further substitutions $z = y + c/2$ and $a = c - c^2/4$, we obtain

$$\dot{z} = z^2 + a,$$

which is the normal form for a *fold bifurcation*.

(iii) Bifurcation diagram is shown in Figure 3.13.

(b) $\dot{x} = rx + x^3 - x^5$

(i) Stationary solutions are roots of

$$rx + x^3 - x^5 = 0,$$

that is, $X_1 = 0$ which exists for all values of r, and four other solutions are roots of

$$r + x^2 - x^4 = 0.$$

After the substitution $y = x^2$, we obtain the quadratic equation

$$r + y - y^2 = 0,$$

for which we only consider solutions with $y > 0$.
For the above quadratic equation, the solutions are

$$Y_{1,2} = \frac{1 \pm \sqrt{1 + 4r}}{2}.$$

They exist for $r > -\frac{1}{4}$ and correspond to two pairs of solutions to the original equation obtained by the back substitution $x = \pm\sqrt{y}$.

When $r = -\frac{1}{4}$ we have $Y_{1,2} = \frac{1}{2}$, which implies

$$X_2 = X_3 = \frac{1}{\sqrt{2}} \quad \text{and} \quad X_4 = X_5 = -\frac{1}{\sqrt{2}}.$$

Thus, at $r = -\frac{1}{4}$, two fold bifurcations take place—one at $x = \frac{1}{\sqrt{2}}$ and one at $x = -\frac{1}{\sqrt{2}}$.

We should remember that only $y = x^2 > 0$ corresponds to solutions of the original equation.

The root

$$Y_1 = \frac{1 + \sqrt{1+4r}}{2}$$

which corresponds to $X_{2,5} = \pm\sqrt{Y_1}$, is always positive. By contrast

$$Y_2 = \frac{1 - \sqrt{1+4r}}{2} > 0$$

corresponding to $X_{3,4} = \pm\sqrt{Y_2}$, is positive only when $r < 0$.

When $r > 0$, the two solutions $X_{3,4} = \pm\sqrt{Y_2}$ disappear. At $r = 0$, we have $Y_2 = 0$, and these disappearing solutions collapse to zero, coinciding with X_1. Thus, at $r = 0$ we have a *pitchfork bifurcation.*

Stability of equilibria is defined by the sign of the derivative

$$f_x(x, r) = r + 3x^2 - 5x^4.$$

For $X_1 = 0, f_x(0, r) = r$, which is negative (X_1 is stable) for $r < 0$ and positive (X_1 is unstable) for $r > 0$.

From this, we can infer the stability of the other four solutions.

In the range $-\frac{1}{4} < r < 0$ we have we have four solutions in addition to $X_1 = 0$:

- X_2 and X_3—both positive, and equal to $1/\sqrt{2}$ when $r = -\frac{1}{4}$;
- X_4 and X_5—both negative, and equal to $-1/\sqrt{2}$ when $r = -\frac{1}{4}$.

The middle pair (X_3 and X_4) merge at $x = 0$ when $r = 0$ and disappear for $r > 0$. Since X_1 is stable when these two exist, they must both be unstable in order to satisfy the conditions for a *subcritical pitchfork bifurcation.*

Consequently, the outer equilibria—$X_2 > \frac{1}{\sqrt{2}}$ and $X_5 < -\frac{1}{\sqrt{2}}$—are stable.

(ii) Near $(x, r) = (0, 0)$, we get the third-order approximation $\dot{x} \approx rx + x^3$, which is the normal form of a pitchfork bifurcation.

Near $r = -1/4$, $x = 1/\sqrt{2}$, where $X_{2,3}$ appear/disappear, make the substitutions $z = x - 1/\sqrt{2}$ and $c = r + 1/4$. In the second-order approximation, we obtain

$$\dot{z} = cz + c/\sqrt{2} - \sqrt{2}z^2.$$

Assuming $c \sim z^2$ (so that cz is of higher order), we may neglect the first term and get

$$\dot{z} = c/\sqrt{2} - \sqrt{2}z^2.$$

With the substitutions $\eta = \sqrt{2}z$ and $\mu = c$, this becomes

$$\dot{\eta} = \mu - \eta^2,$$

which is the normal form for a fold bifurcation.
Near $(x, r) = (-1/\sqrt{2}, -1/4)$, set $z = x + 1/\sqrt{2}$ and $c = r + 1/4$. Further substitutions ($\eta = \sqrt{2}z$ and $\mu = c$) lead to the normal form

$$\dot{\eta} = -\mu + \eta^2.$$

(iii) Bifurcation diagram is shown in Figure 3.14.

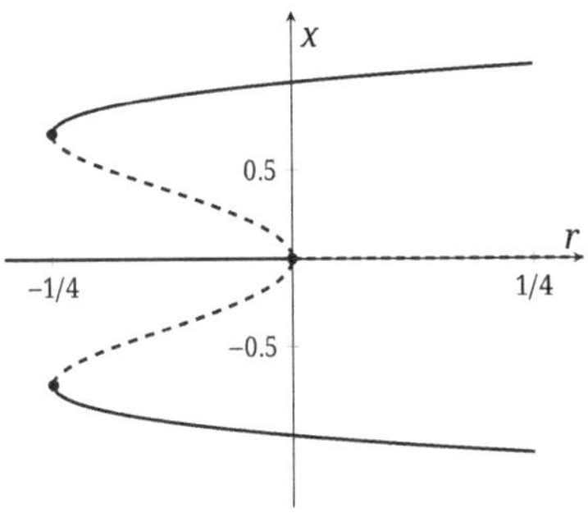

Figure 3.14: Bifurcation diagram for the flow $\dot{x} = rx + x^3 - x^5$. Bifurcation points are marked by dots. The point $(r, x) = (0, 0)$ corresponds to a subcritical pitchfork bifurcation. The points $(r, x) = (-\frac{1}{4}, \frac{1}{\sqrt{2}})$ and $(-\frac{1}{4}, -\frac{1}{\sqrt{2}})$ correspond to fold bifurcations.

4 Multivariable continuous dynamical systems

In the two previous chapters, we considered 1D maps and 1D flows. We saw that while the dynamical regimes in 1D maps can include quite sophisticated patterns, such as oscillations and chaos, those in 1D flows are restricted to monotonic changes of the system. Our next task is to analyze dynamical regimes in continuous systems described by more than one variable (multidimensional flows) and to determine under what conditions we can expect the formation of oscillations and chaotic regimes.

4.1 2D flows

We begin the study of multivariable systems by considering the simplest object of this type, namely, an autonomous two-variable system, or 2D flow, which is represented by two coupled differential equations

$$\begin{cases} \dot{x} = f(x,y), \\ \dot{y} = g(x,y). \end{cases} \tag{4.1}$$

These equations describe a dynamical system whose state is given by two quantities, x and y, that change over time t. The system is autonomous since the functions on the right-hand side of both equations do not explicitly depend on time. Similarly to the requirements we have put for the 1D flows in the previous chapter, we will request that the functions on the right-hand side of both equations have continuous derivatives, i. e., $f(x,y), g(x,y) \in C^1(\Omega \subset \mathbb{R}^2)$. The solution of this system is given by two functions, $x(t)$ and $y(t)$, which transform the given equations into identities

$$\begin{cases} \dot{x}(t) \equiv f(x(t),y(t)), \\ \dot{y}(t) \equiv g(x(t),y(t)). \end{cases}$$

Similarly to the 1D flows considered in the previous chapter, the system (4.1) has a family of solutions. A unique solution is specified by setting the initial conditions

$$\begin{cases} x(t=0) = x_0, \\ y(t=0) = y_0, \end{cases} \tag{4.2}$$

and is graphically represented by two plots, $x(t)$ and $y(t)$. On the other hand, solutions can also be illustrated in the (x,y)-plane. After dividing the second equation in the system (4.1) by the first, we obtain the differential equation

$$\frac{dy}{dx} = \frac{g(x,y)}{f(x,y)} \tag{4.3}$$

https://doi.org/10.1515/9783111348339-004

which does not involve the independent variable t and can be treated as a single differential equation describing the dynamical system, as was done in the previous chapter. That is, the plane (x, y) represents a phase plane, the Eq. (4.3) defines the direction field, and its solutions are represented by curves called *integral curves, phase trajectories*, or *orbits*. The state of the system (4.1) at time t is given by the point $(x(t), y(t))$, and its evolution corresponds to motion along an orbit in the direction of increasing time.

As defined in the Chapter 3 (Definition 3.1), a phase plane with a representative set of phase trajectories is called a *phase portrait*. Commonly, a phase portrait also includes a direction field, represented by a set of vectors tangential to the trajectories and pointing towards the direction of motion.

Example 4.1. Consider the 2D flow

$$\begin{cases} \dot{x} = y, \\ \dot{y} = yx(x-1). \end{cases} \tag{4.4}$$

The solution $x(t), y(t)$ for this system (obtained using Matlab) for the initial conditions $x(t = 0) = 0.25$ and $y(t = 0) = 0.5$ is shown in Figure 4.1A, while its phase portrait (a collection of orbits in the (x, y)-plane) is presented in Figure 4.1B. Although orbits can be determined as solutions of the single differential equation

$$\frac{dy}{dx} = x(x-1),$$

which is obtained by dividing the second equation by the first in system (4.4), the direction of motion along the orbits can only be deduced using the time derivatives from the original system (4.4). As the sign of $\dot{x}$ coincides with the sign of y, the flow is directed along the positive x-axis when $y > 0$ and along the negative x-axis when $y < 0$.

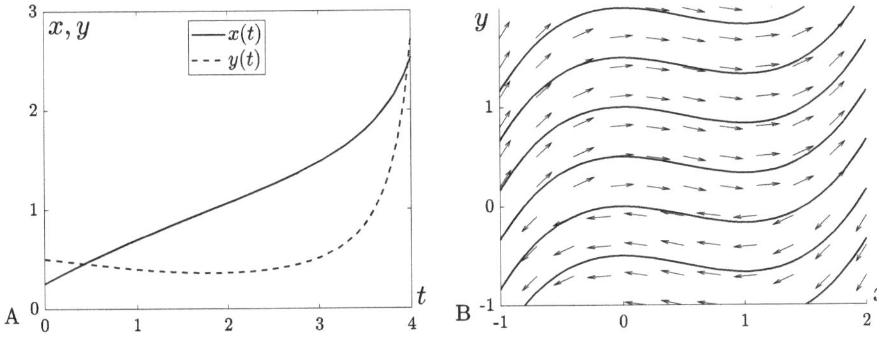

Figure 4.1: Panel A: Solution of the system (4.4), given by the functions $x(t)$ and $y(t)$, for the initial conditions $x_0 = 0.25$ and $y_0 = 0.5$. **Panel B:** Direction field and integral curves for the system (4.4). Solid lines represent phase trajectories, while small vectors (with slope $g(x, y)/f(x, y)$) indicate the direction field.

In the case of 1D flows, the motion was always from left to right in the phase portraits (see Figure 3.1) since the horizontal axis represented the variable t, i. e., time. However, in 2D flows the motion may proceed in either direction (as the horizontal axis here corresponds to the coordinate x) and is determined by the orientation of the vectors representing the direction field of the flow.

Note, that a nonautonomous 1D flow, $\dot{x} = f(x,t)$, can be represented as an autonomous 2D flow by introducing a second variable $y = t$ as follows:

$$\begin{cases} \dot{x} = f(x,y), \\ \dot{y} = 1. \end{cases}$$

In general, a nonautonomous n-variable system can be transformed into an autonomous $(n+1)$-variable system.

Another important note is that a second-order differential equation can be transformed into a 2D flow. For example, in the equation of motion $\ddot{x} = f(x,\dot{x})$, alongside the independent variable x, representing position, we can introduce another variable $\dot{x}$, corresponding to velocity. With the substitution $y = \dot{x}$ we obtain

$$\begin{cases} \dot{x} = y, \\ \dot{y} = f(x,y), \end{cases}$$

which is an autonomous two-variable dynamical system. Similarly, an nth-order differential equation can be transformed into a system of n first-order differential equations or an nD flow.

4.2 Nullclines

As we noted earlier, orbits, as represented on the phase plane, are found as solutions of Eq. (4.3). In particular, the derivative dy/dx defines the slope of an orbit, that is,

$$\tan\alpha = \frac{g(x,y)}{f(x,y)}, \tag{4.5}$$

where α is the angle between the tangent and the horizontal axis. The important cases are $\alpha = 0$ and $\alpha = 90°$, when the tangent is horizontal and vertical, respectively.

Definition 4.1. The set of points on the phase plane, where the slope of the tangent to the orbits is fixed, that is,

$$\frac{g(x,y)}{f(x,y)} = c,$$

where c is a constant, is called an **isocline**.

Example 4.2. Consider the second-order differential equation $\ddot{x}+x=0$, which is known in physics as the *harmonic oscillator*. Introducing the variable $y=\dot{x}$ transforms this equation into a system of first-order differential equations

$$\ddot{x}+x=0 \quad \Rightarrow \quad \begin{cases} \dot{x}=y, \\ \dot{y}=-x. \end{cases} \tag{4.6}$$

Division of the second equation by the first gives

$$\frac{dy}{dx}=-\frac{x}{y},$$

which can be integrated by separation of variables, yielding the family of circles $x^2+y^2=r^2$.

The solution can also be analyzed graphically using isoclines as follows:

1. isocline $\tan\alpha=1$,

$$\frac{dy}{dx}=1 \quad \Rightarrow \quad \frac{-x}{y}=1 \quad \Rightarrow \quad y=-x.$$

All orbits cross the line $y=-x$ at an angle $\alpha=45°$.

2. isocline $\tan\alpha=-1$,

$$\frac{dy}{dx}=-1 \quad \Rightarrow \quad \frac{-x}{y}=-1 \quad \Rightarrow \quad y=x.$$

All orbits cross the line $y=x$ at an angle $\alpha=-45°$.

3. isocline $\tan\alpha=0$,

$$\frac{dy}{dx}=0 \quad \Rightarrow \quad \frac{-x}{y}=0 \quad \Rightarrow \quad x=0.$$

All orbits cross the line $x=0$ horizontally, i. e., at $\alpha=0°$.

4. isocline $\tan\alpha=\infty$,

$$\frac{dy}{dx}=\infty \quad \Rightarrow \quad \frac{-x}{y}\to\infty \quad \Rightarrow \quad y=0.$$

All orbits cross the line $y=0$ vertically, i. e., at $\alpha=90°$.

Plotting these isoclines together with the direction field shows that the orbits rotate around the origin in the phase plane. It is reasonable to assume that the trajectories form circles, as illustrated in Figure 4.2A. This becomes even clearer when more isoclines are added to the phase plane.

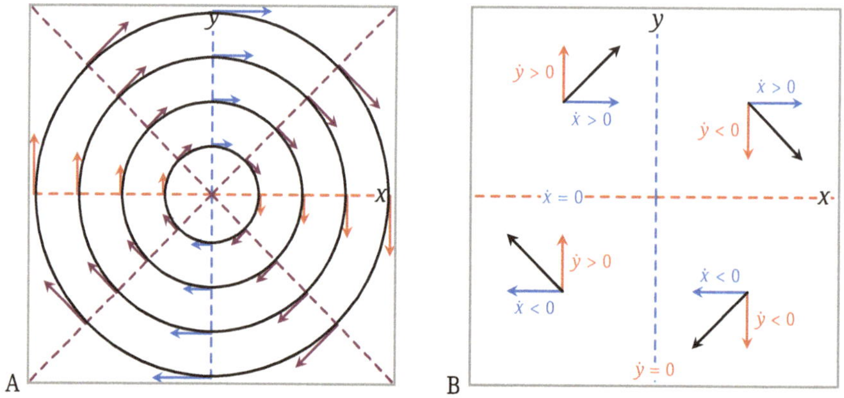

Figure 4.2: Phase portrait and orientation of flows for the harmonic oscillator. Panel A: Isoclines with accompanying direction vectors (colored) and sample orbits (circles) in the phase plane. **Panel B:** The phase plane is divided by the nullclines ($\dot{x} = 0$ given by the dashed red line and $\dot{y} = 0$ given by the dashed blue) of the harmonic oscillator into four domains represented by the quadrants of the plane. In the first quadrant, x is increasing and y is decreasing, leading to flows in the southeast direction. In the second quadrant, both time derivatives are positive, producing flows in the northeast direction. The directions of flows in the remaining two quadrants are also indicated.

Definition 4.2. The isoclines

$$\frac{g(x,y)}{f(x,y)} = c,$$

with $c = 0$ or $c = \infty$, are called **nullclines**.

These are isoclines where the tangents to the orbits are horizontal or vertical. The case $c = \infty$ (see (4.5)) corresponds to $\dot{x} = f(x,y) = 0$, and this isocline is called the *x-nullcline*. At points on this isocline, the tangent to the phase trajectory is vertical ($\tan \alpha = \infty \to \alpha = 90°$).

Similarly, $c = 0$ corresponds to $\dot{y} = g(x,y) = 0$, and this isocline is called the *y-nullcline*. At points on this isocline, the tangent to the phase trajectory is horizontal ($\tan \alpha = 0 \to \alpha = 0°$).

Nullclines are used to identify domains where the flows have qualitatively different directions.

- $\dot{x} = f(x,y) = 0$, x-coordinate is not changing and the tangents to orbits are vertical when they cross x-nullcline.
- $\dot{y} = g(x,y) = 0$, y-coordinate is not changing and the tangents to orbits are horizontal when they cross y-nullcline.
- The derivative $\dot{x}$ is positive on one side of x-nullcline and negative on the other. That is, x-coordinate is increasing along the orbit on one side of x-nullcline and decreasing on the other.

- Similarly, the derivative $\dot{y}$ is positive (i. e., y-coordinate is increasing) on one side of the y-nullcline and negative (i. e., y-coordinate is decreasing) on the other.
- Nullclines divide the phase plane into domains, so that in each domain, the sign of $\dot{x}$ and $\dot{y}$ do not change and therefore, the dynamics of x- and y-coordinates is qualitatively preserved. This makes it possible to estimate the direction of flows in all these domains.

In Example 4.2, we found that the nullclines for harmonic oscillator are lines $x = 0$ and $y = 0$, which divide the phase plane into four domains corresponding to the four quadrants. To determine the directions of the flows in each of these domains, we analyze the system as defined by Eq. (4.6). For instance, in the first quadrant (Q1—top right), $\dot{x} = y > 0$, meaning that x increases along the orbit over time, and $\dot{y} = -x < 0$, meaning that y decreases. This corresponds to flows in the southeast direction. Using similar reasoning we find that in the second quadrant the flows point northeast, in the third quadrant northwest, and in the fourth quadrant southwest. This is illustrated in Figure 4.2B.

4.3 Properties of phase trajectories

As we noted earlier, phase trajectories representing solutions of ordinary differential equations form patterns reminiscent of flows. This analogy is supported by the following properties of phase trajectories:

1. Phase trajectories of the system (4.1) do not cross each other, provided that the derivative

$$\frac{dy}{dx} = \frac{g(x,y)}{f(x,y)}$$

   is defined and the functions $f(x,y)$ and $g(x,y)$ are continuous. This derivative becomes undefined only if $f(x,y) = 0$ and $g(x,y) = 0$ simultaneously, i. e., at points in the phase plane where $\dot{x} = \dot{y} = 0$. Such points are called stationary or equilibrium points and will be studied in detail later. An intuitive explanation of the non-crossing property follows from the fact that the initial conditions (4.2) uniquely determine the solution of (4.1); if two solutions were to cross, then the crossing point could serve as the initial condition for two different solutions, contradicting uniqueness.
2. Moreover, phase trajectories cannot even touch each other. This is a weaker version of the previous property and again follows from uniqueness of solutions for given initial conditions. However, this property may fail if the derivative of either of the functions $f(x,y)$ and $g(x,y)$ is discontinuous.

It follows that a phase trajectory in a two-dimensional autonomous system can intersect either itself or another trajectory only at an equilibrium point. These properties form the basis of the so-called *Poincaré–Bendixson theorem*, which imposes strong restrictions on the types of dynamical regimes that can occur in two-dimensional flows.

Theorem 4.1 (Poincaré–Bendixson Theorem). *If a phase trajectory in a two-dimensional autonomous flow is bounded, then its long-term behavior is one of the following:*
- *It converges to an equilibrium point; or*
- *It converges to a closed orbit.*

We omit the proof of this theorem, which can be found, for example, in [11]. The formulation given here is not fully precise in the strict mathematical sense, but it captures the essential qualitative feature of 2D flows. It states that bounded trajectories in planar systems may be represented by closed orbits, which correspond to periodic solutions (and, as we know from the previous chapter, such periodic solutions cannot occur in 1D flows). Such periodic solutions are known as *limit cycles*. We will study limit cycles in detail later in this chapter; for now, we simply illustrate this concept with the following example.

Example 4.3. Van der Pol oscillator is a dynamical system described by the second-order differential equation

$$\ddot{x} - \mu(1 - x^2)\dot{x} + x = 0,$$

which, using the substitution $y = \dot{x}$, can be written as a system of two first-order differential equations

$$\begin{cases} \dot{x} = y, \\ \dot{y} = \mu(1 - x^2)y - x. \end{cases} \tag{4.7}$$

This represents a modification of the harmonic oscillator (4.6), with an additional term in the second equation corresponding to nonlinear damping. The presence of this term transforms the oscillations of arbitrary amplitude in the harmonic oscillator (see Figure 4.2) into a limit cycle, as illustrated in Figure 4.3.

The existence of limit cycles in 2D flows makes them richer than 1D flows in terms of possible dynamical regimes. As we have shown in the previous chapter, no periodic solutions can exist in 1D autonomous flows. On the other hand, as a corollary of Poincaré–Bendixon theorem, chaos is ruled out in 2D flows. However, as we will see later, chaotic regimes can form in 3D flows.

4.4 Equilibrium points and their stability

We have seen that stationary solutions are the most important solutions for 1D maps and 1D flows. They are also crucial members of the family of solutions in 2D flows.

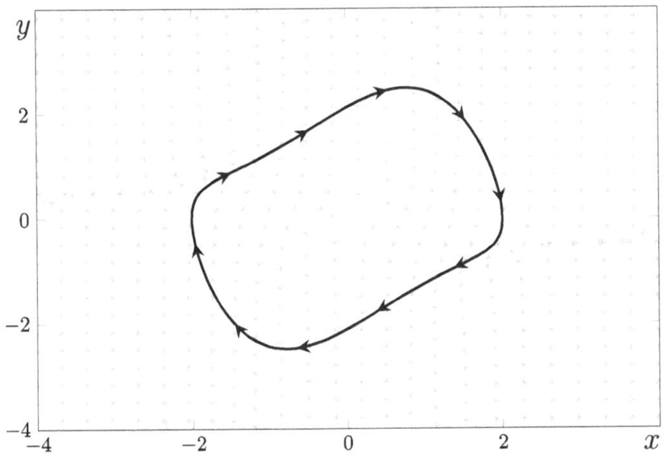

Figure 4.3: Illustration of a stable limit cycle forming in the Van der Pol model (4.7). The direction field (shown by grey arrows) and the limit cycle (shown by the solid looped line with arrows indicating the direction of flow) are obtained for the parameter value $\mu = 0.8$.

Definition 4.3. **Stationary solutions** of the system (4.1) are defined by the conditions

$$\begin{cases} \dot{x} = 0, \\ \dot{y} = 0 \end{cases} \quad \text{or equivalently} \quad \begin{cases} f(x,y) = 0, \\ g(x,y) = 0. \end{cases}$$

Points in the phase plane that satisfy this algebraic system are called *equilibrium* or *stationary points*. As we seen earlier, a key property of an equilibrium point is whether trajectories in its vicinity converge to it or not.

Definition 4.4. An equilibrium point (X, Y) is said to be **asymptotically stable** if there exists $\delta > 0$ such that for any initial condition (x_0, y_0) satisfying

$$(x_0 - X)^2 + (y_0 - Y)^2 < \delta^2,$$

the solution satisfies $x(t) \to X$ and $y(t) \to Y$ as $t \to \infty$. Here, $x_0 = x(0)$ and $y_0 = y(0)$.

Definition 4.5. An equilibrium point (X, Y) is called **stable** if, for every $\varepsilon > 0$, there exists $\delta > 0$ such that for any initial condition (x_0, y_0) satisfying

$$(x_0 - X)^2 + (y_0 - Y)^2 < \delta^2,$$

the corresponding solution $(x(t), y(t))$ satisfies

$$\left(x(t) - X\right)^2 + \left(y(t) - Y\right)^2 < \varepsilon^2$$

for all $t \geq 0$.

Note, that every asymptotically stable equilibrium point is stable, but not every stable equilibrium point is asymptotically stable. In the latter case, trajectories remain close to the equilibrium but do not converge to it as $t \to \infty$.

4.5 Linear 2D flows

Consider the linear two-dimensional flow with constant coefficients

$$\begin{cases} \dot{x} = a_{11}x + a_{12}y, \\ \dot{y} = a_{21}x + a_{22}y. \end{cases} \tag{4.8}$$

The equilibrium points of this flow are given by the roots of the algebraic system

$$\begin{cases} a_{11}x + a_{12}y = 0, \\ a_{21}x + a_{22}y = 0. \end{cases}$$

If these two equations are not equivalent, then the only solution is $x = y = 0$, so that the unique equilibrium point lies at the origin.

Let us rewrite the system (4.8) in matrix form by introducing the vector variable $\mathbf{u}$ and the matrix A

$$\mathbf{u} = \begin{pmatrix} x \\ y \end{pmatrix}, \quad A = \begin{pmatrix} a_{11} & a_{12} \\ a_{21} & a_{22} \end{pmatrix}.$$

With this notation, the system (4.8) takes the compact form

$$\dot{\mathbf{u}} = A\mathbf{u}. \tag{4.9}$$

Theorem 4.2. *If the matrix $A \in \mathbb{R}^{2\times 2}$ has a pair of linearly independent eigenvectors, then the solution of the system (4.9) can be written as*

$$\mathbf{u}(t) = c_1\mathbf{b}_1 e^{\lambda_1 t} + c_2\mathbf{b}_2 e^{\lambda_2 t}, \tag{4.10}$$

where $(\mathbf{b}_1, \lambda_1)$ and $(\mathbf{b}_2, \lambda_2)$ are the eigenvector–eigenvalue pairs of the matrix A, and c_1, c_2 are constants determined by the initial conditions [3].

Proof. Introduce the matrix $B = (\mathbf{b}_1, \mathbf{b}_2)$ and make the substitution $\mathbf{u} = B\mathbf{v}$. Then, the system (4.9) becomes

$$\dot{\mathbf{u}} = A\mathbf{u} \quad \Rightarrow \quad B\dot{\mathbf{v}} = AB\mathbf{v}.$$

Now, compute AB

$$AB = A(\mathbf{b}_1, \mathbf{b}_2) = (A\mathbf{b}_1, A\mathbf{b}_2) = (\lambda_1 \mathbf{b}_1, \lambda_2 \mathbf{b}_2) = (\mathbf{b}_1, \mathbf{b}_2)\begin{pmatrix} \lambda_1 & 0 \\ 0 & \lambda_2 \end{pmatrix} = BD.$$

Thus, $B\dot{\mathbf{v}} = BD\mathbf{v}$, where D is a diagonal matrix, with the diagonal entries given by the eigenvalues of matrix A. Cancelling B (which is invertible since $\mathbf{b}_1$, $\mathbf{b}_2$ are linearly independent) yields

$$\dot{\mathbf{v}} = D\mathbf{v}, \qquad \begin{pmatrix} \dot{v}_1 \\ \dot{v}_2 \end{pmatrix} = \begin{pmatrix} \lambda_1 & 0 \\ 0 & \lambda_2 \end{pmatrix}\begin{pmatrix} v_1 \\ v_2 \end{pmatrix} = \begin{pmatrix} \lambda_1 v_1 \\ \lambda_2 v_2 \end{pmatrix}.$$

The equations for v_1 and v_2 are therefore decoupled and solved independently as follows:

$$\dot{v}_1 = \lambda_1 v_1, \quad \dot{v}_2 = \lambda_2 v_2 \quad \Rightarrow \quad v_1(t) = v_{10} e^{\lambda_1 t}, \quad v_2(t) = v_{20} e^{\lambda_2 t},$$

where $v_{10} = v_1(0)$ and $v_{20} = v_2(0)$.

Returning to $\mathbf{u}$, we obtain

$$\begin{aligned} \mathbf{u}(t) = B\mathbf{v}(t) &= (\mathbf{b}_1, \mathbf{b}_2)\begin{pmatrix} v_{10} e^{\lambda_1 t} \\ v_{20} e^{\lambda_2 t} \end{pmatrix} \\ &= v_{10} \mathbf{b}_1 e^{\lambda_1 t} + v_{20} \mathbf{b}_2 e^{\lambda_2 t}, \end{aligned}$$

which proves the result. □

The following conclusions are drawn from the theorem and its proof:

- The constants c_1 and c_2 in the solution (4.10) are the coordinates of the initial state (x_0, y_0) expressed in the basis formed by the eigenvectors $\mathbf{b}_1$ and $\mathbf{b}_2$ of the matrix A.
- Phase trajectories cannot cross the straight lines through the origin that are parallel to the eigenvectors of A. These lines are often referred to as *eigenlines*.
- The equation for phase trajectories in terms of v_1 and v_2 can be obtained directly as follows:

$$\frac{dv_1}{dv_2} = \frac{\lambda_1 v_1}{\lambda_2 v_2} \quad \Rightarrow \quad \lambda_2 \ln v_1 = \lambda_1 \ln v_2 + c \quad \Rightarrow \quad v_1^{\lambda_2} = C v_2^{\lambda_1},$$

where $C = e^c$ is a constant determined by the initial conditions. This relation shows that phase trajectories are represented by parabolas when the eigenvalues have the same sign and by hyperbolas when their signs are opposite.

The system (4.8) is called *homogeneous* because its right-hand side contains no terms independent of the variables x and y. By contrast, an *inhomogeneous system* has such terms, which, in matrix form, is represented by a constant vector $\mathbf{a}$

$$\dot{\mathbf{u}} = A\mathbf{u} + \mathbf{a}. \tag{4.11}$$

If the matrix A is invertible, we can apply the transformation

$$\dot{\mathbf{u}} = A(\mathbf{u} + A^{-1}\mathbf{a}).$$

Introducing a shifted variable

$$\mathbf{v} = \mathbf{u} + A^{-1}\mathbf{a},$$

we obtain

$$\dot{\mathbf{v}} = A\mathbf{v},$$

which reduces the system (4.11) to a homogeneous flow in terms of $\mathbf{v}$.

Note that $\mathbf{v} = 0$ corresponds to

$$\mathbf{u} = -A^{-1}\mathbf{a},$$

which is the equilibrium point of the inhomogeneous system. In all other respects, the analysis of system (4.11) is identical to that of the homogeneous system (4.9).

4.6 Types of equilibrium points

The dynamical regimes described by the solution (4.10) differ significantly depending on whether the eigenvalues λ_1 and λ_2 are real or complex. We begin by examining the case in which the eigenvalues are real.

4.6.1 Real eigenvalues

Assume that the initial conditions are chosen so that $c_1 = 0$. In this case, the solution reduces to

$$\mathbf{u}(t) = c_2\mathbf{b}_2 e^{\lambda_2 t},$$

which remains parallel to the eigenvector $\mathbf{b}_2$ at all times. On the phase plane, the state of the system is therefore represented by a point constrained to lie on the line through the origin (*eigenline*) parallel to $\mathbf{b}_2$. If $\lambda_2 > 0$, this point moves away from the origin, while if $\lambda_2 < 0$, it approaches the origin.

Similarly, if $c_2 = 0$, the trajectory lies along the *eigenline* parallel to $\mathbf{b}_1$, diverging to infinity when $\lambda_1 > 0$ or converging to the origin when $\lambda_1 < 0$.

In the general case, when both c_1 and c_2 are nonzero, the solution is a superposition of the two components, producing a trajectory that either converges to the origin or

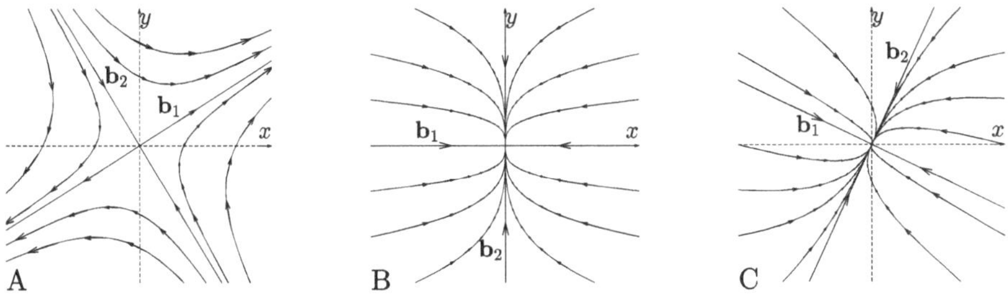

Figure 4.4: Phase portraits of linear systems with real eigenvalues. Panel A: Eigenvalues of opposite site correspond to the equilibrium called *saddle*. $\lambda_1 > 0$, and trajectories along the vector $\mathbf{b}_1$ diverge to infinity. $\lambda_2 < 0$, and trajectories along the vector $\mathbf{b}_2$ converge to the origin. Phase trajectories have a hyperbolic shape tending to infinity and approaching the eigenline parallel to $\mathbf{b}_1$ when $t \to \infty$ and that parallel to $\mathbf{b}_2$ when $t \to -\infty$. **Panels B–C:** Both eigenvalues are negative and all phase trajectories tend to the equilibrium point $(0,0)$, which is called *stable node*. In Panel B eigenlines coincide with axis, while in Panel C they don't. Both these cases can be considered as illustrations to solutions to the Example 4.4.

diverges to infinity, depending on the signs of λ_1 and λ_2. Illustrations of phase portraits of systems with real eigenvalues is given in Figure 4.4.

Provided that both eigenvectors are nonzero, there are three possible dynamical regimes corresponding to the following three types of equilibria at $(0,0)$:

1. If $\lambda_1 < 0$ and $\lambda_2 < 0$, all phase trajectories tend towards the equilibrium point $(0,0)$. The origin is therefore an *asymptotically stable* equilibrium and is referred to as a *stable node* or *sink* (see Figure 4.4B–C).
2. If $\lambda_1 < 0$ and $\lambda_2 > 0$, trajectories do not generally converge to the origin, although those with $c_2 = 0$ (along the eigenline parallel to the stable eigenvector) do (see Figure 4.4A). The equilibrium point is *unstable* and is called a *saddle*.
3. If $\lambda_1 > 0$ and $\lambda_2 > 0$, all trajectories (except the trivial one starting at the origin) diverge to infinity. The equilibrium point is *unstable* and is called an *unstable node* or *source*.

Phase trajectories have hyperbolic shapes in case of a saddle and parabolic—in case of nodes.

Example 4.4. Consider two blocks of material, with initial temperatures T_{10} and T_{20}, placed in an environment maintained at a constant temperature T_s. We aim to describe the dynamics of the system, namely, the evolution of the block temperatures over time, arising from heat exchange between the blocks and their surroundings.

Solution:

The state of the system is described by two variables, $T_1(t)$ and $T_2(t)$, representing the temperatures of the blocks. As we have seen earlier (see Example 1.10 in Chapter 1), heat transfer leads to changes in the temperature according to Newton's law of cooling.

In the simplest case, the blocks are thermally isolated from each other (i. e., they do not touch and no heat is exchanged between them). The governing system of equations is then

$$\begin{cases} \dot{T}_1 = -k_1(T_1 - T_s), \\ \dot{T}_2 = -k_2(T_2 - T_s), \end{cases} \tag{4.12}$$

where k_1 and k_2 are positive constants that determine the rates of temperature change for each block, and T_{10}, T_{20} provide the initial conditions. Since the equations are uncoupled, they can be solved independently. The solutions show that the temperature of each block converges exponentially to the ambient temperature T_s:

$$T_1(t) = (T_{10} - T_s)e^{-k_1 t} + T_s,$$
$$T_2(t) = (T_{20} - T_s)e^{-k_2 t} + T_s.$$

Alternatively, the flow (4.12) can be interpreted as a linear inhomogeneous system of the form (4.11), with the equilibrium located at (T_s, T_s). In this case, the matrix A is diagonal with eigenvalues $-k_1$ and $-k_2$, so the equilibrium point is a *stable node*, and its eigenvectors are aligned with the coordinate axes of the phase plane.

Phase trajectories of the system can be obtained by dividing the first equation of (4.12) by the second

$$\frac{dT_1}{dT_2} = \frac{k_1(T_1 - T_s)}{k_2(T_2 - T_s)}.$$

Integration of this equation (and taking into account initial conditions) gives

$$|T_1 - T_s|^{k_2} = C|T_2 - T_s|^{k_1}, \quad \text{where } C = \frac{|T_{10} - T_s|^{k_2}}{|T_{20} - T_s|^{k_1}}.$$

Assuming $T_1 > T_s$ and $T_2 > T_s$, the absolute values can be dropped, yielding for the explicit solution

$$T_1(T_2) = T_s + (T_{10} - T_s)\left(\frac{T_2 - T_s}{T_{20} - T_s}\right)^{k_1/k_2}.$$

The corresponding phase portrait of the system is shown in Figure 4.4B.

In the more general case, the blocks can be touching each other requiring that the heat transfer between them should also be taken into account as follows:

$$\begin{cases} \dot{T}_1 = -k_1(T_1 - T_s) - k_3(T_1 - T_2), \\ \dot{T}_2 = -k_2(T_2 - T_s) - k_4(T_2 - T_1), \end{cases} \tag{4.13}$$

where positive constants k_3 and k_4 define the rate of temperature change of two blocks due to the heat transfer between them. The system is now described by coupled, inhomogeneous linear equations. It can be reduced to a homogeneous form by introducing the shifted variables $u_1 = T_1 - T_s$ and $u_2 = T_2 - T_s$, which measure deviations of the block temperatures from the surrounding temperature. In terms of new variables, we have

$$\begin{cases} \dot{u}_1 = -(k_1 + k_3)u_1 + k_3 u_2, \\ \dot{u}_2 = k_4 u_1 - (k_2 + k_4)u_2, \end{cases}$$

The system can be presented (see Eq. (4.9)) as $\dot{\mathbf{u}} = A\mathbf{u}$, where

$$\mathbf{u} = \begin{pmatrix} u_1 \\ u_2 \end{pmatrix}, \quad A = \begin{pmatrix} -(k_1 + k_3) & k_3 \\ k_4 & -(k_2 + k_4) \end{pmatrix},$$

and we can use Eq. (4.10) to find phase trajectories. Illustration of typical phase portrait for this system is given in Figure 4.4C.

The more subtle case arises for both kinds of nodes when the eigenvalues coincide, $\lambda_1 = \lambda_2 (= \lambda)$:

- If the matrix A possesses two linearly independent eigenvectors, then the phase trajectories, as given by the equation (4.10), represented by a straight lines, and the equilibrium is called a *star node*. This is illustrated in Figure 4.5A.

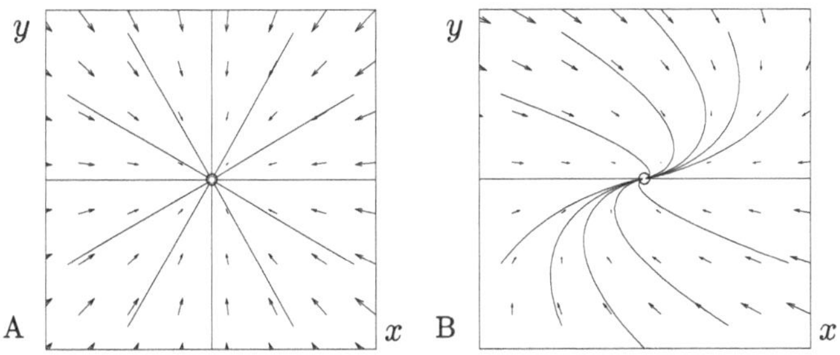

Figure 4.5: Case when $\lambda_1 = \lambda_2 (< 0)$. **Panel A:** If there exist two linearly independent eigenvectors the trajectories are not curved and the equilibrium is called a *star*. **Panel B:** If there is only one eigenvector, then phase trajectories are curved and tend to become parallel to the eigenvalue as time tends to (plus or minus) infinity. The equilibrium is called an *improper node*.

- If there is only one linearly independent eigenvector (say, $\mathbf{b}$), then the solution (4.10) should be corrected and takes the form

$$\mathbf{u}(t) = c_1 \mathbf{b} e^{\lambda t} + c_2 (\mathbf{b}t + \tilde{\mathbf{b}}) e^{\lambda t}, \tag{4.14}$$

where $\tilde{\mathbf{b}} = (A - \lambda I) \cdot \mathbf{b}$. In this case, the equilibrium is referred to as an *improper node* (also called a *degenerate sink/source*). It is illustrated in Figure 4.5B.

A special situation arises when matrix A is not invertible and one or both its eigenvalues are equal to zero:

- *One eigenvalue equal to zero, the other nonzero.* In this case, one of the two terms in the solution (4.10) is constant, and the system possesses a *line of stationary solutions* along the eigenline corresponding to $\lambda = 0$. Along this line, there is no motion, while trajectories in the direction associated with the nonzero eigenvalue either converge to the line (if $\lambda < 0$) or diverge away from it (if $\lambda > 0$).
- *Both eigenvalues equal to zero.* If the matrix A has two linearly independent eigenvectors corresponding to $\lambda = 0$, then every point in the phase plane is an equilibrium point. If, however, A has only one eigenvector, then the solution (4.14) reduces to

$$\mathbf{u}(t) = c_1\mathbf{b} + c_2(\mathbf{b}t + \tilde{\mathbf{b}}),$$

which describes motion with constant velocity along the line parallel to the eigenvector $\mathbf{b}$.

4.6.2 Complex eigenvalues

Now, let us consider the case when the matrix A has the complex eigenvalues

$$\lambda_{1,2} = \alpha \pm i\beta, \tag{4.15}$$

with corresponding complex eigenvectors represented as

$$\mathbf{b}_{1,2} = \mathbf{b}_\mathrm{r} \pm i\mathbf{b}_\mathrm{im},$$

where $\mathbf{b}_\mathrm{r}$ and $\mathbf{b}_\mathrm{im}$ are real vectors. In this situation, the solution (4.10) can be rewritten using Euler's formula as

$$\mathbf{u}(t) = c_1(\mathbf{b}_\mathrm{r} + i\mathbf{b}_\mathrm{im})e^{\alpha t}(\cos\beta t - i\sin\beta t) + c_2(\mathbf{b}_\mathrm{r} - i\mathbf{b}_\mathrm{im})e^{\alpha t}(\cos\beta t + i\sin\beta t),$$

where c_1 and c_2 are complex constants determined by the initial conditions. If we regroup terms and use the fact that c_1 and c_2 can be chosen to make the solution real, we obtain

$$\mathbf{u}(t) = \tilde{c}_1e^{\alpha t}(\mathbf{b}_\mathrm{r}\cos\beta t - \mathbf{b}_\mathrm{im}\sin\beta t) + \tilde{c}_2e^{\alpha t}(\mathbf{b}_\mathrm{r}\sin\beta t + \mathbf{b}_\mathrm{im}\cos\beta t),$$

where $\tilde{c}_1$ and $\tilde{c}_2$ are real constants determined by the initial conditions.

This solution can be represented in matrix form as

$$\mathbf{u}(t) = e^{\alpha t}(\mathbf{b}_\mathrm{r}, \mathbf{b}_\mathrm{im})\begin{pmatrix} \cos\beta t & \sin\beta t \\ -\sin\beta t & \cos\beta t \end{pmatrix}\begin{pmatrix} \tilde{c}_1 \\ \tilde{c}_2 \end{pmatrix}.$$

The obtained formula allows the following interpretation:
- The matrix composed of oscillatory functions $\cos(\beta t)$ and $\sin(\beta t)$ reflects the rotational nature of the motion along phase trajectory with angular frequency β.
- The exponential factor $e^{\alpha t}$ determines whether trajectories spiral outward ($\alpha > 0$), inward ($\alpha < 0$), or remain closed ($\alpha = 0$).
- The vectors $\mathbf{b}_\mathrm{r}$ and $\mathbf{b}_\mathrm{im}$ span the phase plane and determine the shape of the trajectories, i. e., elliptic or circular (if they are orthogonal and of equal length).

Correspondingly, in the case when matrix A has complex eigenvalues (4.15), we have the following three types of equilibria:
1. If the real part of the eigenvalues is negative ($\alpha < 0$), then the equilibrium is called a *stable focus, stable spiral* or *spiral sink*.
2. If the real part of the eigenvalues is positive ($\alpha > 0$), then the equilibrium is called an *unstable focus, unstable spiral* or *spiral source*.
3. If the real part of the eigenvalues is equal to zero ($\alpha = 0$), then the equilibrium is called a *center*.

Example 4.5. The output power of refrigerator's radiator, x, and its internal temperature, y, are related according to the following equations:

$$\begin{cases} \dot{x} = y - T_0, \\ \dot{y} = k(T_s - y) - x, \end{cases} \tag{4.16}$$

where T_0 is a target internal temperature and T_s is the external (room) temperature. The parameter k characterises the insulation of the fridge and represents the rate of heat transfer between the room and the refrigerator (in accordance with Newton's law of cooling). Describe the dynamics of x and y for the case $k = 1$.

Solution:
This is an inhomogeneous linear system with the equilibrium determined from the conditions $\dot{x} = 0$ and $\dot{y} = 0$. From the first equation, we obtain $Y = T_0$, and substituting it into the second equation, we get $X = k(T_s - T_0)$. Therefore, the equilibrium point is

$$(X, Y) = \big(k(T_s - T_0), T_0\big),$$

which indicates that the radiator output x settles at the value $k(T_s - T_0)$, to maintain the internal temperature at its target level T_0.

The matrix A defining this linear system is

$$A = \begin{pmatrix} 0 & 1 \\ -1 & -k \end{pmatrix}.$$

Its eigenvalues are obtained from the characteristic equation

$$\lambda^2 + k\lambda + 1 = 0,$$

which yields

$$\lambda_{1,2} = \frac{-k \pm \sqrt{k^2 - 4}}{2}.$$

For $k = 1$, the eigenvalues are complex with negative real parts, and therefore, the equilibrium point correspond to a stable spiral (see Figure 4.6A). The real part, $\alpha = -0.5$, determines the rate of exponential decay (dumping), while the imaginary part, $\beta = \sqrt{3}/2$, gives the angular frequency of oscillation.

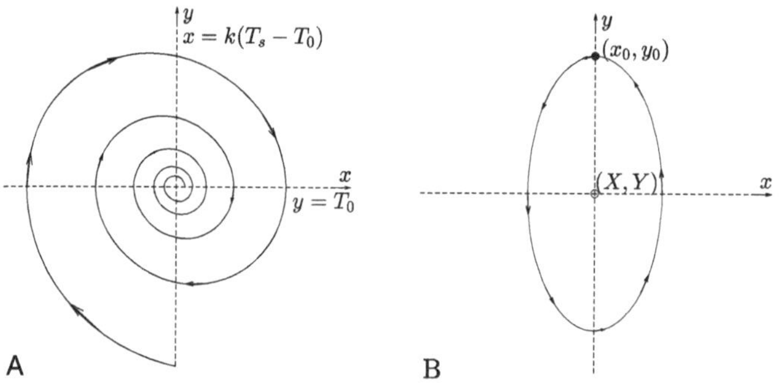

Figure 4.6: Phase portraits: spiral and ellipse. Panel A: Phase trajectory corresponding to a stable focus, illustrating the solution of Example 4.5. **Panel B:** Phase trajectory corresponding to a centre, illustrating the solution of Example 4.6.

Example 4.6. The average number per unit area of plants, $x(t)$, depends on the average number per unit area of animal species, $y(t)$ because this species eats plants. The numbers are assumed to satisfy (for x and y both positive) the differential equations

$$\begin{cases} \dot{x} = 1 - y, \\ \dot{y} = 4(x - 1). \end{cases} \tag{4.17}$$

Find the phase trajectory representing the solution of the system for the initial conditions, $x(0) = 1, y(0) = 1.5$.

Solution:

The equations describing the dynamics of the considered populations form an inhomogeneous linear system. The equilibrium is found from the condition $\dot{x} = \dot{y} = 0$, which yields

$$X = Y = 1.$$

Phase trajectories are found from the equation obtained by dividing the second equation by the first as follows:

$$\frac{\dot{y}}{\dot{x}} = \frac{dy}{dx} = \frac{4(x-1)}{1-y}, \quad \text{or} \quad (1-y)dy = 4(x-1)dx.$$

The solution of this differential equation is given by the algebraic equation

$$4(x-1)^2 + (y-1)^2 = C^2,$$

where the constant C is found from the initial conditions, i. e.,

$$C^2 = 4(x(0)-1)^2 + (Y(0)-1)^2 = 0.25, \quad \text{or} \quad C = 0.5$$

Thus, the phase trajectory is given by the ellipse, which is shown in Figure 4.6B. The direction of rotation along this ellipse can be found in the following way. At the initial point $x(0) = 1, y(0) = 1.5$, which is the topmost point of the ellipse, we have $\dot{x} = -0.5 < 0$ and $\dot{y} = 0$, i. e., the direction vector is horizontal, pointing to the left. From this, we conclude that the dynamics is given by counterclockwise rotation along the found ellipse.

4.7 Poincaré diagram

As the type of equilibrium in (4.9) is determined by the eigenvalues of matrix A, we can express the conditions for observing various types of equilibria in this system in terms of the trace and determinant of A.

Indeed, the eigenvalues of matrix A are the roots of the characteristic equation [3]

$$\begin{vmatrix} a_{11}-\lambda & a_{12} \\ a_{21} & a_{22}-\lambda \end{vmatrix} = 0,$$

or equivalently,

$$\lambda^2 - (a_{11}+a_{22})\lambda + (a_{11}a_{22} - a_{12}a_{21}) = 0.$$

This can be written in terms of the trace and determinant of matrix A as

$$\lambda^2 - \operatorname{tr} A \cdot \lambda + \det A = 0,$$

and therefore,

$$\lambda_{1,2} = \frac{\operatorname{tr} A \pm \sqrt{\operatorname{tr}^2 A - 4\det A}}{2}. \tag{4.18}$$

From this formula, we can derive the conditions corresponding to the various types of equilibria.

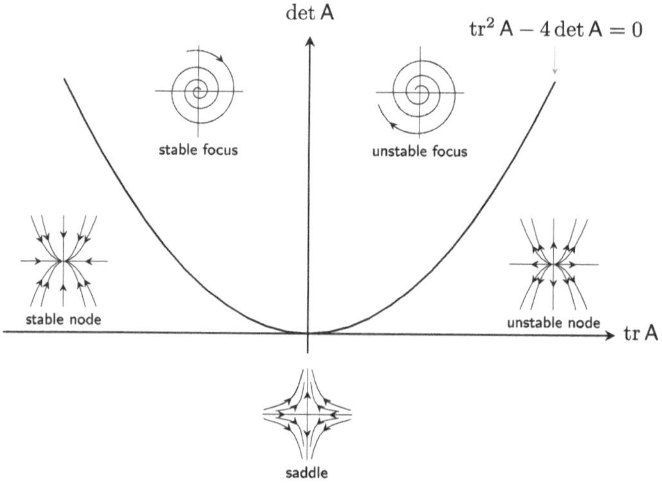

Figure 4.7: Poincaré diagram. Classification of equilibrium points for the linear system $\dot{\mathbf{u}} = A\mathbf{u}$ in the $(\det A, \operatorname{tr} A)$ plane.

Note that both eigenvalues are real if $\operatorname{tr}^2 A \geq 4 \det A$. Under this condition, we can observe the following:

1. If $\operatorname{tr} A < 0$ and $\det A > 0$, both eigenvalues are real and negative, the equilibrium is a *stable node*.
2. If $\operatorname{tr} A < 0$ and $\det A < 0$, the system has two real eigenvalues of opposite signs—the equilibrium is a *saddle*.
3. If $\operatorname{tr} A > 0$ and $\det A > 0$, both eigenvalues are real and positive—the equilibrium is an *unstable node*.

   If $\operatorname{tr}^2 A < 4 \det A$, then the eigenvalues are complex conjugates, and the sign of their real part is determined by the sign of $\operatorname{tr} A$. In this case,
4. If $\operatorname{tr} A > 0$, the equilibrium point is an *unstable spiral.*
5. If $\operatorname{tr} A < 0$, the equilibrium point is a *stable spiral.*
6. If $\operatorname{tr} A = 0$, the equilibrium point is a *centre.*

All these results are summarised in the Poincaré diagram shown in Figure 4.7.

4.8 Analysis of nonlinear systems

In the previous sections, we have identified dynamical regimes that can be observed in linear systems. However, many natural phenomena exhibit more sophisticated regimes, indicating that they are essentially nonlinear and should be described by more general equations, i. e., Eq. (4.1).

In Chapter 3, we considered the equation describing the population dynamics of a single species, $\dot{x} = r(x)x$, where x is the population size and $r(x)$ defines its growth rate. If $r(x)$ is a constant, $r(x) = r_0$, then we obtain the Malthusian model, which is linear and predicts exponential growth of the population. If $r(x)$ is a linear function of x, $r(x) = r_0 - x/K$, we obtain the Verhulst model, which involves quadratic nonlinearity and predicts saturation of the population size.

For a two-species system, where the size of the second population is given by the variable y, one can assume that the rate of population growth will depend on the sizes of both populations, $\dot{x} = r(x,y)x$. In the simplest case, when $r(x,y)$ is a linear function of both variables, we obtain the following system that involves quadratic nonlinearity:

$$\begin{cases} \dot{x} = (r_1 + b_1 x + b_2 y)x, \\ \dot{y} = (r_2 + b_3 x + b_4 y)y. \end{cases} \tag{4.19}$$

Here, parameters b_2 and b_3 define the relationship between species. If $b_2 = b_3 = 0$, the species do not interact and each follows its own logistic equation. If both parameters are negative, they reduce each other's growth rate, which corresponds to *competition* between the two species. If both parameters are positive, the species enhance each other's growth, which corresponds to *symbiosis* or *mutualism*. Finally, if the parameters have opposite signs, the species exhibit a *predator–prey* relationship.

As we have seen, to analyze the dynamics of the system (4.19) qualitatively, we determine the nullclines defined by $\dot{x} = 0$ and $\dot{y} = 0$. These curves partition the phase plane into regions, where the direction of the flow differs qualitatively. In this way, we sketch the *direction field*, and obtain a qualitative picture of the system's dynamics in multiple regions of phase plane.

We also know that equilibrium points for the system (4.1) are obtained as the roots of the algebraic system

$$\begin{cases} f(x,y) = 0, \\ g(x,y) = 0, \end{cases}$$

or as points of intersection of the x- and y-nullclines. This implies that the quadratic system (4.19) has up to four equilibrium points, provided that the x-nullclines and y-nullclines do not coincide (in which case, the number of equilibria becomes infinite).

Example 4.7 (Battle model). Two armies, X and Y, consisting of x_0 and y_0 soldiers, respectively, fight until one or the other army has no soldiers left. Assume that the army sizes, $x(t)$ and $y(t)$, during the battle are described by the differential equations

$$\begin{cases} \dot{x} = -axy, \\ \dot{y} = -bxy, \end{cases} \tag{4.20}$$

where the parameters $a > 0$ and $b > 0$ define the effectiveness of each army in killing enemy soldiers. What is an outcome of the battle?

Solution:
Nullcline $\dot{x} = 0$ is given by the lines $x = 0$ and $y = 0$. Nullcline $\dot{y} = 0$ is given by the same lines. Thus, we have infinite number of equilibrium points, represented by all points on horizontal and vertical axes. This is in line with the expectation that the battle stops when either $x = 0$ or $y = 0$.

As the variables x and y are both positive, we are restricted to considering solutions of the model in the first quadrant, where, according to (4.20), $\dot{x} < 0$ and $\dot{y} < 0$. The equation for phase trajectories is obtained by dividing one equation of the system by the other as follows:

$$\frac{dy}{dx} = \frac{b}{a}.$$

Integration of this equation yields an implicit expression for the trajectories, with the integration constant determined by the initial conditions

$$y = \frac{b}{a}x + \left(y_0 - \frac{b}{a}x_0\right).$$

This, depending on the initial sizes of the armies, leads to the following three possible outcomes:

- If $y_0 - \frac{b}{a}x_0 > 0$, army Y wins ($y > 0$ when $x = 0$).
- If $y_0 - \frac{b}{a}x_0 = 0$, it is a draw (both armies vanish simultaneously).
- If $y_0 - \frac{b}{a}x_0 < 0$, army X wins ($x > 0$ when $y = 0$).

The corresponding phase trajectories are shown in Figure 4.8A.

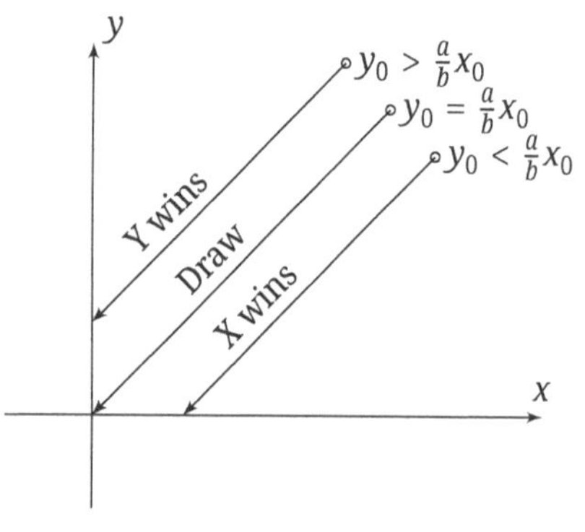

Figure 4.8: Phase portrait of the system (4.20), with phase trajectories illustrating three possible outcomes of the battle.

In this example, the system has infinitely many equilibria. Typically, nonlinear systems possess only a few such points. To determine the stability (and type) of equilibrium points in a nonlinear system, the system is linearized in the vicinity of each equilibrium, and the eigenvalues of the resulting linear system are calculated.

Linearization is achieved using the first-order Taylor expansion. If (X, Y) is an equilibrium point, then we can write $f(x,y) = f(X + \epsilon, Y + \delta)$, where ϵ and δ are small perturbations: $|\epsilon| \ll 1$ and $|\delta| \ll 1$. Applying the first-order Taylor expansion gives

$$f(x,y) = f(X,Y) + f_x'(X,Y)\epsilon + f_y'(X,Y)\delta = f_x'(X,Y)\epsilon + f_y'(X,Y)\delta$$

since $f(X, Y) = 0$ at the equilibrium. Similarly,

$$g(x,y) = g_x'(X,Y)\epsilon + g_y'(X,Y)\delta.$$

We note that $\dot{x} = \dot{\epsilon}$ and $\dot{y} = \dot{\delta}$ since the differences between the two pairs of variables are given by the constants X and Y. Thus, linearization of the original system at the equilibrium point (X, Y) yields the following system:

$$\begin{cases} \dot{\epsilon} = f_x'(X,Y), \epsilon + f_y'(X,Y), \delta, \\ \dot{\delta} = g_x'(X,Y), \epsilon + g_y'(X,Y), \delta, \end{cases}$$

or, in matrix form,

$$\begin{pmatrix} \dot{\epsilon} \\ \dot{\delta} \end{pmatrix} = J(X,Y) \begin{pmatrix} \epsilon \\ \delta \end{pmatrix},$$

where the matrix $J(X, Y)$ is called the **Jacobian matrix** and defined as

$$J(X,Y) = \begin{pmatrix} f_x'(X,Y) & f_y'(X,Y) \\ g_x'(X,Y) & g_y'(X,Y) \end{pmatrix}.$$

The type of the equilibrium at (X, Y) is determined by the eigenvalues of the Jacobian matrix $J(X, Y)$, in exactly the same way as the equilibrium at $(0, 0)$ was classified by the eigenvalues of matrix A for the linear system (4.9).

Example 4.8 (Predator–prey model). Consider the dynamics of two animal species, X and Y, given by the following equations:

$$\begin{cases} \frac{dx}{dt} = ax - bxy, \\ \frac{dy}{dt} = -cy + dxy, \end{cases} \tag{4.21}$$

where a, b, c and d are positive constants and $x(t)$ and $y(t)$ denote the sizes of the species. The system (4.21) is a well-known Lotka–Volterra predator–prey model; species X is a prey (i. e., hare), whose size would grow exponentially (term ax) if not hunted by the predator (term $-bxy$). Species Y is a predator (i. e., lynx), which would die out exponentially (term $-cy$) if it did not consume the prey (term dxy). Analyze the population dynamics described by this system.

Solution:
The x-nullcline is found from the equation $ax - bxy = 0$ and is given by two lines, $x = 0$ and $y = a/b$.

The y-nullcline is found from the equation $-cy + dxy = 0$ and is given by two lines $y = 0$ and $x = c/d$.

We note that we are interested in solutions located in the first quadrant ($x > 0$ and $y > 0$), which is divided by the nullclines $y = a/b$ and $x = c/d$ into four domains, with $\dot{x} > 0$ if $y < a/b$ and $\dot{y} > 0$ if $x < c/d$.

Equilibrium points are found as points of intersection of the nullclines. The nullclines $x = 0$ and $y = 0$ intersect at $(X_1, Y_1) = (0, 0)$, and the nullclines $y = a/b$ and $x = c/d$, intersect at $(X_2, Y_2) = (c/d, a/b)$.

To analyze the stability of equilibrium points, we find the Jacobian

$$J = \begin{pmatrix} a - by & -bx \\ dy & dx - c \end{pmatrix}.$$

The Jacobian at the first equilibrium is

$$J(0, 0) = \begin{pmatrix} a & 0 \\ 0 & -c \end{pmatrix}.$$

The eigenvalues at this equilibrium are roots of $\begin{vmatrix} a-\lambda & 0 \\ 0 & -c-\lambda \end{vmatrix} = 0$. They are found as $\lambda_1 = a$ and $\lambda_2 = -c$, which implies that the equilibrium $(X_1, Y_1) = (0, 0)$ is a saddle.

Similarly, the Jacobian at the second equilibrium

$$J\left(\frac{c}{d}, \frac{a}{b}\right) = \begin{pmatrix} a - b\frac{a}{b} & -b\frac{c}{d} \\ d\frac{a}{b} & d\frac{c}{d} - c \end{pmatrix} = \begin{pmatrix} 0 & -b\frac{c}{d} \\ d\frac{a}{b} & 0 \end{pmatrix}.$$

The eigenvalues of the Jacobian

$$\begin{vmatrix} -\lambda & -b\frac{c}{d} \\ d\frac{a}{b} & -\lambda \end{vmatrix} = \lambda^2 + ca = 0 \quad \Rightarrow \quad \lambda_{1,2} = \pm i\sqrt{ca}.$$

Thus, the second equilibrium $(X_2, Y_2) = (c/d, a/b)$ is a center, and, therefore, the phase trajectories are represented by closed loops.

For the system (4.21), phase trajectories can be found analytically. Divide one of the equations by the other and separate variables as follows:

$$\frac{dx}{dy} = \frac{(a - by)x}{(dx - c)y} \quad \Rightarrow \quad \left(d - \frac{c}{x}\right)dx = \left(\frac{a}{y} - b\right)dy.$$

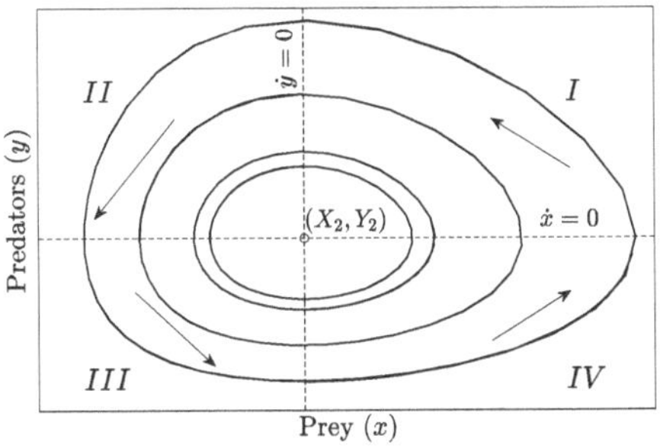

Figure 4.9: Nullclines, direction field, and phase trajectories for the system (4.21). Solid lines represent phase trajectories, dashed lines represent nullclines, and vectors qualitatively indicate the direction field in each of the four domains defined by the nullclines.

Both sides of this differential equation can be integrated as follows:

$$\int \left(d - \frac{c}{x}\right) dx = dx - c \ln x + c_1,$$
$$\int \left(\frac{a}{y} - b\right) dy = a \ln y - by + c_2.$$

Thus,

$$dx + by - \ln(x^c y^a) = C \tag{4.22}$$

is an equation for phase trajectories, with constant C found from the initial conditions. The phase portrait for the Lotka–Volterra system (4.20) is shown in Figure 4.9.

Thus, the Lotka–Volterra predator–prey model describes an oscillating system, with the amplitude of oscillations defined by the initial conditions. Linearization of the system at the equilibrium $(X_2, Y_2) = (c/d, a/b)$ indicates that the phase trajectories in the vicinity of this point are approximately elliptic. However, the exact shape of the phase trajectories, as found analytically (4.22) and shown in Figure 4.9, is not elliptic.

4.9 Saddle point and its stable and unstable manifolds

Linearization of the system at an equilibrium point captures the system's dynamics in a small neighborhood of this point. A global picture, however, can either be obtained from the exact solution (as in the case of the predator–prey system (4.21)) or derived qualitatively through analysis of the location and types of equilibria in the given system. This analysis is heavily based on the identification of so-called *stable* and *unstable manifolds* of saddle points.

Assume that one of the equilibria in the considered system is a saddle, i. e., one of the eigenvalues of the Jacobian calculated at this equilibrium point is negative and the other is positive. In a linear system, phase orbits along one of the eigenlines converge to the saddle. In a nonlinear system, this eigenline is not necessarily a straight line, but there are still phase trajectories converging to the saddle along it.

Definition 4.6. A collection of points forming phase trajectories that converge to a saddle is called a **stable manifold.**

In other words, the *stable manifold* is the collection of all initial points such that the corresponding phase trajectories converge to a saddle. In a linear system, it is represented by the eigenline corresponding to the negative eigenvalue (see Figure 4.4A).

There are also orbits that converge to the saddle point in the inverse flow, i. e., in the system with inverted time, where the evolution is associated with $t \to -\infty$.

Definition 4.7. A collection of points forming phase trajectories that converge to a saddle in an inverse-time flow is called an **unstable manifold.**

Stable and unstable manifolds are illustrated in Figure 4.10. Stable manifolds act as repellers, that is, nearby orbits diverge from them. Similarly, unstable manifolds act as attractors, as nearby orbits converge to them.

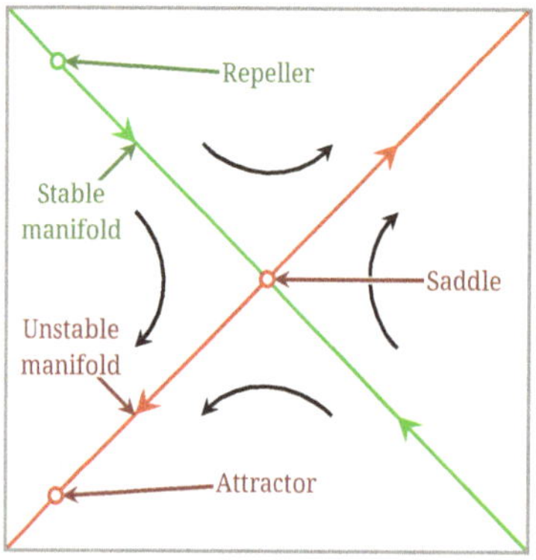

Figure 4.10: Illustration of the stable and unstable manifolds of the saddle, the direction of flows in the regions they form, and the possible locations of an attractor and a repeller.

If there is an attractor (i. e., a stable fixed point) and a saddle point in the system, the attractor will always be located on the unstable manifold of the saddle point. It follows that the unstable manifold is an orbit that starts at the saddle and either goes to infinity or terminates at an attractor. Similarly, if the system has a saddle point and a repeller (i. e., an unstable node), then the repeller is located on the stable manifold of the saddle, in which case the stable manifold is an orbit starting at the repeller and terminating at the saddle. So, we have identified orbits connecting the saddle with the attractor or the repeller.

Definition 4.8. An orbit connecting two equilibrium points is called a **heteroclinic orbit.**

There is also a possibility that the unstable manifold coincides with the stable manifold, in which case it represents an orbit that starts at the saddle and terminates at the same saddle.

Definition 4.9. An orbit connecting an equilibrium point with itself is called a **homoclinic orbit.**

Stable manifolds act as borders between basins of attraction of different attractors and, for this reason, are called *separatrices*. All the previous observations concerning the properties of stable and unstable manifolds make possible qualitative analysis of nonlinear systems that cannot be solved analytically. Such analysis involves the following three steps:
1. Finding the nullclines and sketching the direction of flow in the system.
2. Finding all equilibria and determining their types.
3. Drawing the qualitative phase portrait of the system.

Example 4.9 (Lotka–Volterra competition model). Perform the qualitative analysis of the dynamical system described by the following equations:

$$\begin{cases} \dot{x} = (4 - x - y)x, \\ \dot{y} = (6 - y - 2x)y. \end{cases} \tag{4.23}$$

Solution:
We note that if the variables x and y are considered as the sizes of two populations, the system (4.23) describes the competitive dynamics between these species.

Nullclines
x-nullcline, $\dot{x} = 0$, is given by two straight lines: $x = 0$ and $y = 4 - x$. $\dot{x} > 0$ when either $x > 0$ and $y < 4 - x$, or $x < 0$ and $y > 4 - x$, and $\dot{x} < 0$ elsewhere in the plane.

y-nullcline, $\dot{y} = 0$, is represented by the straight lines $y = 0$ and $y = 6 - 2x$. $\dot{y} > 0$ when either $y > 0$ and $y < 6 - 2x$, or $y < 0$ and $y > 6 - 2x$, and is negative otherwise. Nullclines are shown in Figure 4.11A.

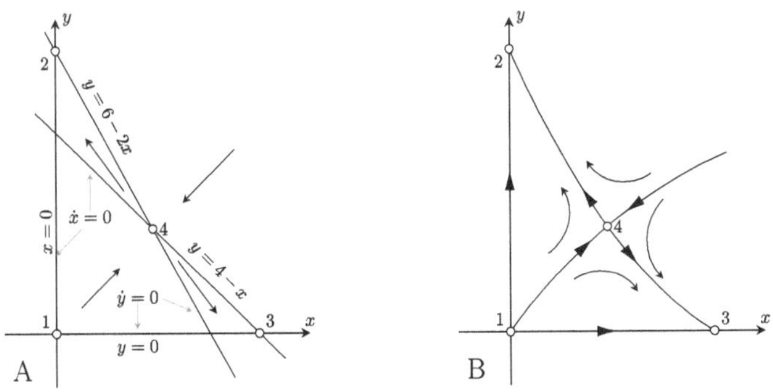

Figure 4.11: Panel A: Nullclines and direction field for the system (4.23). **Panel B:** Sketch of flows for this system. Equilibria are indicated and numbered as follows: 1—unstable node, 2 and 3—stable nodes, 4—saddle.

Equilibria

Equilibrium points are found at the intersections of the x-nullclines and the y-nullclines. There are four such points $(X_1, Y_1) = (0, 0)$, $(X_2, Y_2) = (0, 6)$, $(X_3, Y_3) = (4, 0)$, and $(X_4, Y_4) = (2, 2)$. Equilibria are shown in Figure 4.11A.

Stability and type of equilibrium points are defined by the eigenvalues of the Jacobian, which is found to be

$$J = \begin{pmatrix} 4 - 2x - y & -x \\ -2y & 6 - 2y - 2x \end{pmatrix}.$$

Now we find the Jacobian and its eigenvalues at the fixed points.

$$J(0, 0) = \begin{pmatrix} 4 & 0 \\ 0 & 6 \end{pmatrix},$$

with eigenvalues 4 and 6, thus (X_1, Y_1) is an unstable node.

$$J(0, 6) = \begin{pmatrix} -2 & 0 \\ -12 & -6 \end{pmatrix},$$

with eigenvalues −2 and −6, thus (X_2, Y_2) is a stable node.

$$J(4, 0) = \begin{pmatrix} -4 & -4 \\ 0 & -2 \end{pmatrix},$$

with eigenvalues −2 and −4, thus (X_3, Y_3) is a stable node.

$$J(2, 2) = \begin{pmatrix} -2 & -2 \\ -4 & -2 \end{pmatrix},$$

with $\lambda_{1,2} = -2 \pm 2\sqrt{2}$. Since one of these eigenvalues is positive and the other is negative, the fixed point (X_4, Y_4) is a saddle.

Sketch of the phase portrait

To draw the phase portrait of the system, we first plot the equilibrium points and identify the stable and unstable manifolds of the saddle by connecting equilibria with lines representing heteroclinic orbits. The flows are then sketched so that their directions are consistent with those of the heteroclinic orbits, as shown in Figure 4.11B.

It is observed that the unstable manifold of the saddle divides the phase plane into the following two domains: trajectories above this manifold converge to the stable node at $(0, 6)$, while those below converge to the stable node at $(4, 0)$. This means that, depending on the initial population sizes, the system evolves in such a way that only one species survives, while the other becomes extinct.

4.10 2D flows with parameter

Autonomous two-dimensional flow with a single parameter is represented by the system

$$\begin{cases} \dot{x} = f(x,y,c) \\ \dot{y} = g(x,y,c), \end{cases} \tag{4.24}$$

where $f(x,y,c)$ and $g(x,y,c)$ are differentiable functions with respect to the variables x and y and the parameter c. In previous chapters, we have already considered systems with parameters and noted that variation in a parameter affects the location, number, and stability of the equilibria.

Analysis of systems with parameters is focused on the identification of so-called *bifurcation values* of parameters, which correspond to a change in the number of equilibria or in their stability. We have identified three types of bifurcations in 1D flows, namely, fold, transcritical, and pitchfork bifurcations. All of them are also observed in 2D flows.

Example 4.10. Find the bifurcation points in the following 2D flow with parameter μ:

$$\begin{cases} \dot{x} = \mu - x^2 \\ \dot{y} = -y. \end{cases}$$

Solution:
This system has two equilibrium points $P_{1,2} = (\pm\sqrt{\mu}, 0)$, which exist only when $\mu > 0$. Thus, at $\mu = 0$, we have a fold bifurcation. In two-dimensional systems. this is usually called a *saddle–node bifurcation*, as one of the equilibrium points is always saddle and the other is a node (either stable or unstable). To find out who is a saddle and who is a node, we have to find the eigenvalues of the Jacobian

$$J(x,y) = \begin{pmatrix} -2x & 0 \\ 0 & -1 \end{pmatrix}.$$

Now,

$$J(\sqrt{\mu}, 0) = \begin{pmatrix} -2\sqrt{\mu} & 0 \\ 0 & -1 \end{pmatrix},$$

for which $\lambda_1 = -2\sqrt{\mu} < 0$ and $\lambda_2 = -1 < 0$. Thus, $P_1 = (\sqrt{\mu}, 0)$ is a stable node.

Similarly,

$$J(-\sqrt{\mu}, 0) = \begin{pmatrix} 2\sqrt{\mu} & 0 \\ 0 & -1 \end{pmatrix},$$

for which $\lambda_1 = 2\sqrt{\mu} > 0$ and $\lambda_2 = -1 < 0$. Thus, $P_2 = (-\sqrt{\mu}, 0)$ is a saddle.

The bifurcation diagram shown in Figure 3.8C can be regarded as that for this system, expressed in terms of the variable x and the parameter μ.

4.11 Hopf bifurcation

The types of bifurcations that can occur in 1D flows (saddle-node, transcritical, and pitchfork) also occur in 2D flows when one of the eigenvalues of the Jacobian at the equilibrium point becomes equal to zero. In terms of the variables used in the Poincaré diagram (see Figure 4.7), these bifurcations take place when the determinant of the Jacobian at the equilibrium point is zero. This corresponds to the horizontal axis in the Poincaré diagram, which separates saddles (below the horizontal axis) from nodes (unstable nodes in the first quadrant and stable nodes in the second).

There is one additional bifurcation observed in 2D flows that is associated with the transition between stable and unstable spirals. It is called the Hopf, or Andronov–Hopf, bifurcation. According to the Poincaré diagram (Figure 4.7), it takes place when $\det J(X,Y) > 0$ and $\operatorname{tr} J(X,Y) = 0$. Thus, the transition between stable and unstable spirals is associated with a bifurcation at the point (X,Y,C) if the eigenvalues

$$\lambda_{1,2} = \alpha(x,y,c) \pm \beta(x,y,c)$$

of the Jacobian $J(x,y,c)$ at this point satisfy the following conditions:
1. $\alpha(X,Y,C) = 0$ and $\beta(X,Y,C) \neq 0$.
2. $\frac{\partial \alpha}{\partial c}|_{X,Y,C} \neq 0$.

Under certain (nondegeneracy) conditions, the Hopf bifurcation is associated with the formation (or disappearance) of limit cycles. The formulation of the non-degeneracy conditions is given in terms of the so-called first *Lyapunov exponent*

$$\begin{aligned} L = {} & \frac{1}{16}(f_{xxx} + f_{xyy} + g_{xxy} + g_{yyy}) \\ & + \frac{1}{16\beta}[f_{xy}(f_{xx} + f_{yy}) - g_{xy}(g_{xx} + g_{yy}) - f_{xx}g_{xx} + f_{yy}g_{yy}], \end{aligned}$$

which is defined for sufficiently smooth flow-generating functions, that is, $f(x,y,c)$, $g(x,y,c) \in C^{3,3,1}(\Omega \subset \mathbb{R}^3)$. The nondegeneracy condition is
3. $L \neq 0$.

The derivation of this condition is lengthy and can be found in [14]. One can state that these conditions are met almost always in nonlinear 2D flows.

There are two subclasses of Hopf bifurcation. One is the *subcritical Hopf bifurcation*, which takes place when the transition from an unstable spiral to a stable one is accompanied by the formation of an unstable limit cycle. The other is the *supercritical Hopf bifurcation*, which takes place when the transition from a stable spiral to an unstable one is accompanied by the formation of a stable limit cycle.

Example 4.11. Consider the following flow with parameter μ:

$$\begin{cases} \dot{x} = -y + x[\mu - (x^2 + y^2)] \\ \dot{y} = x + y[\mu - (x^2 + y^2)] \end{cases} \tag{4.25}$$

We cannot easily find the nullclines of this system and therefore shall pursue a different approach to gain an idea of the phase trajectories. For example, we can evaluate the flow at some representative points, bearing in mind that the horizontal and vertical components of the flow at the point (x, y) are given by $f(x, y)$ and $g(x, y)$, respectively.

Let us find the components of the flow velocity at four points, $(1, 0)$, $(0, 1)$, $(-1, 0)$, and $(0, -1)$, lying on the unit circle in the phase plane. The results of these calculations are shown in Table 4.1. From this table, we conclude that when $\mu < 1$, the vectors representing the flow velocity point inwards towards the unit circle (see Figure 4.12A). When $\mu > 1$, these vectors point outwards from the unit circle (see Figure 4.12C). When $\mu = 1$, the flow appears to rotate along the unit circle (see Figure 4.12B).

Table 4.1: Components of the flow at four points: $(x, y) = (1, 0), (0, 1), (-1, 0)$, and $(0, -1)$.

x	y	$f(x, y)$	$g(x, y)$
1	0	$\mu - 1$	1
0	1	-1	$\mu - 1$
-1	0	$1 - \mu$	-1
0	-1	1	$1 - \mu$

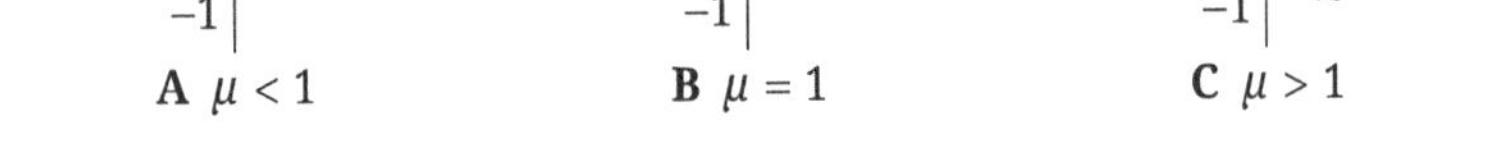

Figure 4.12: Directions of the flow at four representative points for the system (4.25). Panel A: $\mu < 1$, and the flow points inward toward the unit circle. **Panel B:** $\mu = 1$, and the flow is directed tangentially along the unit circle. **Panel C:** $\mu > 1$, and the flow points outward from the unit circle.

It is easy to check that the system (4.25) has only one equilibrium point at $(0,0)$. The Jacobian of the system is

$$J(x,y) = \begin{pmatrix} \mu - 3x^2 - y^2 & -1 \\ 1 & \mu - x^2 - 3y^2 \end{pmatrix},$$

which, at the equilibrium $(0,0)$, becomes

$$J(0,0) = \begin{pmatrix} \mu & -1 \\ 1 & \mu \end{pmatrix}.$$

The eigenvalues of $J(0,0)$ are $\lambda_{1,2} = \mu \pm i$, and therefore the equilibrium at $(0,0)$ is a stable spiral if $\mu < 0$ and an unstable spiral if $\mu > 0$. At $\mu = 0$ we have a (nondegenerate) Hopf bifurcation associated with the existence of a limit cycle. The system (4.25) allows this limit cycle to be found analytically.

Let us replace the variables x and y by their polar counterparts $r = \sqrt{x^2 + y^2}$ and $\theta = \arctan(y/x)$. From

$$\dot{r^2} = \frac{d(x^2 + y^2)}{dt} = 2x\dot{x} + 2y\dot{y} = 2(x^2 + y^2)[\mu - (x^2 + y^2)] = 2r^2(\mu - r^2)$$

and

$$\dot{r^2} = 2r\dot{r},$$

we obtain

$$\dot{r} = r(\mu - r^2). \tag{4.26}$$

This is a first-order differential equation in the normal form for a supercritical pitchfork bifurcation (see Eq. (3.21) and Figure 3.10C) and can be analyzed using techniques presented in the previous chapter. If $\mu < 0$, the only equilibrium point is $r = 0$, which is stable; therefore, $r \to 0$ in the course of time. If $\mu > 0$, then $r = 0$ is unstable, while there appear two additional equilibria, $r = \pm\sqrt{\mu}$, both of which are stable. As r is a position vector, it cannot be negative, and only one of these two equilibria, $r = \sqrt{\mu}$, is meaningful. Thus, if $\mu > 0$, then $r \to \sqrt{\mu}$ in the course of time.

Now, taking into account that

$$\dot{\theta} = \frac{d}{dt}\left(\arctan\frac{y}{x}\right),$$

we get

$$\dot{\theta} = \frac{1}{1 + (y/x)^2}\frac{d}{dt}\left(\frac{y}{x}\right) = \frac{x^2}{x^2 + y^2} \cdot \frac{x\dot{y} - y\dot{x}}{x^2} = \frac{x\dot{y} - y\dot{x}}{x^2 + y^2} = \cdots = 1.$$

Thus, we have $\dot{\theta} = 1$, which means that the point (x, y) representing the state of the system (4.25) rotates counterclockwise around the origin with a constant angular speed. Consequently, the circle of radius $\sqrt{\mu}$ centered at the origin is a stable limit cycle. Therefore, in the system (4.25), we observe a supercritical Hopf bifurcation: When the parameter μ changes from negative to positive, the spiral transforms from stable to unstable, and this transition is accompanied by the formation of a stable limit cycle.

We can also find the analytical solution of the system (4.25) since Eq. (4.26) can be integrated using separation of variables and the method of partial fractions

$$\dot{r} = r(\mu - r^2) \quad \Rightarrow \quad \frac{dr}{r(\mu - r^2)} = dt \quad \Rightarrow$$

$$\frac{1}{\mu}\int\left(\frac{1}{r} + \frac{r}{\mu - r^2}\right)dr = dt \quad \Rightarrow \quad \ln\frac{r^2}{|\mu - r^2|} = 2\mu t + c.$$

From the last equation, we can find $r(t)$ explicitly and show that the radius vector of the state of the system evolves according to

$$r^2 = \frac{A\mu}{A + \exp(-2\mu t)},$$

where $A = \frac{r_0}{\mu - r_0}$ and r_0 is the radius vector corresponding to the initial state of the system. This solution indicates that $r^2 \to \mu$ as $t \to \infty$.

Equation (4.25) represents a normal form for the supercritical Hopf bifurcation. The normal form for the subcritical Hopf bifurcation is given by the system similar to (4.25) but with positive signs inside the square brackets

$$\begin{cases} \dot{x} = -y + x[\mu + (x^2 + y^2)], \\ \dot{y} = x + y[\mu + (x^2 + y^2)]. \end{cases} \tag{4.27}$$

The differential equation for the radius-vector r in this case is

$$\dot{r} = r(\mu + r^2),$$

which is in a normal form for subcritical pitchfork bifurcation (see Figure 3.10A). In this system, when the parameter μ changes from positive to negative, the spiral transforms from unstable to stable, and this transition is accompanied by the formation of an unstable limit cycle. This behavior is the hallmark of a subcritical Hopf bifurcation.

It is not always possible for any flow represented by (4.1) to determine analytically whether it has a limit cycle, as we did for (4.25). In such cases, the following reasoning based on the use of the Poincaré–Bendixson theorem often proves helpful. If we can identify a domain such that any phase trajectory inside this domain does not leave it and there are no stable equilibria within the domain, then there must be a limit cycle inside the domain.

This approach can be illustrated using the previous example. On the one hand, the system (4.25) has the only equilibrium $(0,0)$, which is unstable if $\mu > 0$. On the other hand, we have found that if $\mu < 1$, then $\dot{r} < 0$ on the circle $r = 1$ (as shown by Eq. (4.26) and illustrated in Figure 4.12). This means that if $0 < \mu < 1$, phase trajectories inside the unit circle do not leave it and should converge to a stable solution. Since there are no stable equilibria within this domain when $\mu > 0$, the only remaining option is that the phase trajectories inside the unit circle converge to a stable limit cycle.

4.12 Types of dynamical systems

If the flow is smooth (for example, defined by the equations in (4.1) where the functions $f(x,y)$ and $g(x,y)$ are differentiable), then points covering a certain area will flow together so that this area shifts and deforms over time (illustrated schematically in Figure 4.13A). One of the important questions concerning the observed dynamics is whether the size of the area formed by a given collection of points is increasing or decreasing over time.

If the area is increasing, then nearby trajectories diverge, and we have sensitivity to initial conditions, which, as we recall from Chapter 2, was a necessary condition for chaotic behavior in one-dimensional maps. In two-dimensional flows, however, chaotic dynamics cannot occur (due to restrictions imposed by the Poincaré–Bendixson theorem), and therefore we would expect solutions to be unbounded. This means that flows forming an area-expanding domain do not converge to any local attractor.

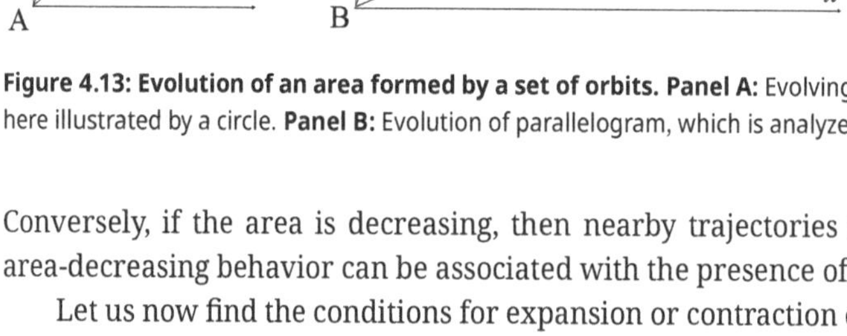

Figure 4.13: Evolution of an area formed by a set of orbits. Panel A: Evolving area of a general shape, here illustrated by a circle. **Panel B:** Evolution of parallelogram, which is analyzed in the text.

Conversely, if the area is decreasing, then nearby trajectories may converge so that area-decreasing behavior can be associated with the presence of an attractor.

Let us now find the conditions for expansion or contraction of an area formed by a certain collection of points in the flow (4.1). To do this, we consider the evolution of an infinitesimal parallelogram (illustrated in Figure 4.13B). The result obtained can then be extended to the area of any shape since any area can be partitioned into a number of small rectangular domains.

Note that the area of a parallelogram built on vectors $\boldsymbol{\delta} = (\delta_x, \delta_y)$ and $\boldsymbol{\epsilon} = (\epsilon_x, \epsilon_y)$ is

$$A = \delta_x \epsilon_y + \delta_y \epsilon_x. \tag{4.28}$$

In particular, if $\delta_x = 0$ and $\epsilon_y = 0$, we obtain a rectangle with sides parallel to the x- and y-axes and area $A = \delta_y \epsilon_x$. Points inside the parallelogram built on $\boldsymbol{\delta}$ and $\boldsymbol{\epsilon}$ when subjected to the system (4.1), will move so that the entire parallelogram shifts and deforms. After an infinitesimal time dt, it can still be considered (to first order) as a parallelogram, though built on different vectors and having a different area. The rate of change of the area of the parallelogram can be found based on the rate of change of the components of vectors $\boldsymbol{\delta}$ and $\boldsymbol{\epsilon}$ as follows:

$$\frac{dA}{dt} = \frac{d(\delta_x \epsilon_y + \delta_y \epsilon_x)}{dt} = \delta_x \frac{d\epsilon_y}{dt} + \frac{d\delta_x}{dt}\epsilon_y + \delta_y \frac{d\epsilon_x}{dt} + \frac{d\delta_y}{dt}\epsilon_x. \tag{4.29}$$

Let one of the vertices of the parallelogram be located at (x_0, y_0), with two adjacent vertices at (x_1, y_1) and (x_2, y_2). In other words, we define the vectors $\boldsymbol{\delta}$ and $\boldsymbol{\epsilon}$ so that

$$\epsilon_x = x_1 - x_0, \quad \epsilon_y = y_1 - y_0, \quad \delta_x = x_2 - x_0, \quad \delta_y = y_2 - y_0.$$

Then, according to Eq. (4.28), the area of the parallelogram is

$$A = (x_2 - x_0)(y_1 - y_0) + (y_2 - y_0)(x_1 - x_0).$$

Getting back to Eq. (4.29), we have

$$\begin{aligned}
\frac{dA}{dt} &= (x_2 - x_0)\frac{d(y_1 - y_0)}{dt} + \frac{d(x_2 - x_0)}{dt}(y_1 - y_0) \\
&\quad + (y_2 - y_0)\frac{d(x_1 - x_0)}{dt} + \frac{d(y_2 - y_0)}{dt}(x_1 - x_0) \\
&= (x_2 - x_0)(\dot{y}_1 - \dot{y}_0) + (\dot{x}_2 - \dot{x}_0)(y_1 - y_0) \\
&\quad + (y_2 - y_0)(\dot{x}_1 - \dot{x}_0) + (\dot{y}_2 - \dot{y}_0)(x_1 - x_0).
\end{aligned}$$

Taking into account that $\dot{x} = f(x,y)$ and $\dot{y} = g(x,y)$, we get

$$\begin{aligned}
\dot{A} &= (x_2 - x_0)(g(x_0, y_1) - g(x_0, y_0)) + (f(x_2, y_0) - f(x_0, y_0))(y_1 - y_0) \\
&\quad + (y_2 - y_0)(f(x_1, y_0) - f(x_0, y_0)) + (g(x_0, y_2) - g(x_0, y_0))(x_1 - x_0).
\end{aligned}$$

Using the first-order Taylor expansion, this becomes

$$\begin{aligned}
\dot{A} &= (x_2 - x_0)\, g'_y(x_0, y_0)(y_1 - y_0) + f'_x(x_0, y_0)(x_2 - x_0)(y_1 - y_0) \\
&\quad + (y_2 - y_0)\, f'_x(x_0, y_0)(x_1 - x_0) + g'_y(x_0, y_0)(y_2 - y_0)(x_1 - x_0).
\end{aligned}$$

Factoring terms, we obtain

$$\begin{aligned}\dot{A} &= [(x_2 - x_0)(y_1 - y_0) + (y_2 - y_0)(x_1 - x_0)](f'_x(x_0, y_0) + g'_y(x_0, y_0)) \\ &= A\,(f'_x(x_0, y_0) + g'_y(x_0, y_0)) = A \cdot \operatorname{tr} J(x_0, y_0).\end{aligned}$$

Thus, the rate of change of the parallelogram's area located at some point (x, y) is defined by the trace of the Jacobian at this point $\operatorname{tr} J(x, y)$ as follows:
1. If $\operatorname{tr} J(x, y) > 0$, the area expands, the system is called *area-expanding*, and phase trajectories starting in the vicinity of (x, y) diverge.
2. If $\operatorname{tr} J(x, y) < 0$, the area contracts, the system is called *dissipative*, and phase trajectories in the vicinity of (x, y) may converge. The term *dissipative* refers to mechanical systems, where the presence of friction results in dissipation of mechanical energy and convergence of phase trajectories to the stable equilibrium.
3. If $\operatorname{tr} J(x, y) = 0$, the area is preserved and the system is called *conservative*. The term *conservative* refers to mechanical systems without friction (such as the harmonic oscillator (4.6)), where the mechanical energy is conserved and motion (i. e., oscillations) continues indefinitely.

As it follows from Eq. (4.18), the trace of a matrix is equal to the sum of its eigenvalues. This implies that
- A linear system with an unstable node or spiral is area-expanding since the sum of eigenvalues in these cases is positive.
- A linear system with a stable node or spiral is dissipative since the sum of eigenvalues in these cases is negative.
- A linear system with a centre (circle) is conservative since the sum of eigenvalues in this case is zero.

The subtle case is a linear system with a saddle, where the sum of eigenvalues can be positive, negative, or zero. This shows that a dissipative system is not necessarily associated with the presence of an attractor, whereas an area-expanding system always contains a repeller.

In linear systems, the Jacobian, $J(x, y)$, is constant over the entire phase plane and, therefore, defining the type of dynamical system is straightforward. In nonlinear systems the Jacobian, $J(x, y)$, depends on the coordinates of the point and, correspondingly, its trace can vary from point to point. However, in many important dynamical systems the sign of the trace, $\operatorname{tr} J(x, y)$, is still the same at all points of the phase plane, and we can therefore determine the type of the system by examining its Jacobian. In general, the system can exhibit different dynamical regimes along different orbits depending on their location in the phase plane. Analysis of the dynamics along a particular orbit involves calculation of Lyapunov exponents (which was introduced in Chapter 2) and can be quite sophisticated. We will return to the calculation of Lyapunov exponents in Chapter 5.

Note that Eq. (4.1) can be written in a vector form as

$$\dot{\mathbf{u}} = \mathbf{F}(x, y),$$

where

$$\mathbf{u} = \begin{pmatrix} x \\ y \end{pmatrix} \quad \text{and} \quad \mathbf{F}(x,y) = \begin{pmatrix} f(x,y) \\ g(x,y) \end{pmatrix}.$$

In this notation, $\operatorname{tr} J(x,y)$ can be represented as the divergence of the vector field $\mathbf{F}(x,y)$, that is,

$$\nabla \cdot \mathbf{F}(x,y) = \frac{\partial f(x,y)}{\partial x} + \frac{\partial g(x,y)}{\partial y}.$$

4.13 3D flows (autonomous 3-variable systems)

Autonomous 3D flows are dynamical systems described by three quantities, say x, y, and z, whose evolution over time t is given by the following three coupled differential equations:

$$\begin{cases} \dot{x} = f(x,y,z), \\ \dot{y} = g(x,y,z), \\ \dot{z} = h(x,y,z). \end{cases} \tag{4.30}$$

As in the case of 1D and 2D flows, we require the functions on the right-hand side to have continuous derivatives, i. e., $f(x,y,z), g(x,y,z), h(x,y,z) \in C^1(\Omega \subset \mathbb{R}^3)$. The solution of this system is given by functions $x(t)$, $y(t)$ and $z(t)$, which transform the equations in the above system into the identities

$$\begin{cases} \dot{x}(t) \equiv f(x(t), y(t), z(t)), \\ \dot{y}(t) \equiv g(x(t), y(t), z(t)), \\ \dot{z}(t) \equiv h(x(t), y(t), z(t)). \end{cases}$$

Theory concerning equilibria in 3D flows is essentially the same as for 2D flows. Equilibria are found as the roots of the algebraic system

$$\begin{cases} f(x,y,z) = 0, \\ g(x,y,z) = 0, \\ h(x,y,z) = 0. \end{cases}$$

To define the type of equilibrium at the point (X, Y, Z), we find the Jacobian at the point

$$J(X,Y,Z)=\begin{pmatrix} f'_x(X,Y,Z) & f'_y(X,Y,Z) & f'_z(X,Y,Z)\\ g'_x(X,Y,Z) & g'_y(X,Y,Z) & g'_z(X,Y,Z)\\ h'_x(X,Y,Z) & h'_y(X,Y,Z) & h'_z(X,Y,Z)\end{pmatrix}. \tag{4.31}$$

We then calculate its eigenvalues. As the Jacobian is now a 3×3 matrix, we have to deal with three eigenvalues, which may all be real or consist of one real and two complex conjugates.

In the case when all eigenvalues of the Jacobian are real, we have the following types of equilibria:

1. If all eigenvalues are negative, the equilibrium is a *stable node.*
2. If all eigenvalues are positive, the equilibrium is an *unstable node.*
3. If there are eigenvalues of opposite signs, the equilibrium is a *saddle point*. If two eigenvalues are positive and one is negative, the saddle has two unstable and one stable manifold, meaning that phase trajectories diverge in a plane and converge along a line. If one eigenvalue is positive and two are negative, then there is one unstable and two stable manifolds, meaning that there is divergence along a line and convergence in a plane.

   In the case when one eigenvalue of the Jacobian is real and the other two are complex conjugates, we have the following types of equilibria:
4. If the sign of the real eigenvalue and the sign of the real part of the complex eigenvalues are negative, the equilibrium is a *stable spiral node*. Phase trajectories spiral in and converge along a line to the equilibrium point.
5. If the sign of the real eigenvalue and the sign of the real part of the complex eigenvalues are positive, the equilibrium is an *unstable spiral node*. Phase trajectories spiral out and diverge along a line from the equilibrium point.
6. If the sign of the real eigenvalue and the sign of the real part of the complex eigenvalues are opposite, the equilibrium is a *saddle-spiral*, and phase trajectories are represented by conical spirals. If the real eigenvalue is positive, then phase trajectories diverge from the equilibrium along a line and spiral in towards this line. If the real eigenvalue is negative, then phase trajectories converge along a line and spiral out from this line.

It follows that the stable equilibrium point is either a node or a spiral node. The condition for the stability of an equilibrium point can be formulated in terms of the Routh–Hurwitz stability criterion. The characteristic equation for the Jacobian (4.31) is given by the polynomial

$$\lambda^3+a_1\lambda^2+a_2\lambda+a_3=0, \tag{4.32}$$

where the coefficients a_1, a_2, and a_3 are defined by the entries of the Jacobian (4.31): $a_1=-\operatorname{tr}J$, $a_3=-\det J$, and $a_2=M_{11}+M_{22}+M_{33}$, which is the sum of the principal minors of J. The Routh–Hurwitz criterion states that if

$$\begin{cases} a_1, a_2, a_3 > 0, \\ a_3 - a_1 a_2 < 0, \end{cases} \tag{4.33}$$

then the real roots of (4.32) are negative and the complex roots have negative real parts, and therefore the steady state is stable.

Limit cycles in 3D flows can have more sophisticated shapes compared to those in 2D flows. For example, limit cycles in 3D space can be doubly periodic. This can be illustrated by an orbit located on the surface of a torus, rotating both around its central axis (of radius R) and around the circular axis (of radius r). The formula describing the surface of a torus in Cartesian coordinates (x, y, z), with its center at the origin and its axis along the z-coordinate, is given by the equations

$$\begin{cases} x(\phi, \theta) = (R + r\cos\phi)\sin\theta, \\ y(\phi, \theta) = (R + r\cos\phi)\cos\theta, \\ z(\phi, \theta) = r\sin\phi, \end{cases} \tag{4.34}$$

where ϕ is the angle around the tube and θ the angle around the central axis (see Figure 4.14). Rotation around the tube can be described by the differential equation $\dot{\phi} = \omega_1$ (uniform rotation), while rotation around the z-axis is given by $\dot{\theta} = \omega_2$. If $\omega_1 = 0$, we have an orbit represented by a circle of radius R with its center at the origin of the coordinate system. If $\omega_2 = 0$, then the orbit is again a circle, but with radius r, centred at some point on the tube axis. If neither ω_1 nor ω_2 is zero, then the orbit representing the limit cycle is helical. Finally, if the ratio ω_1/ω_2 is a rational number, the orbit is closed and the motion along it is periodic. However, if this ratio is irrational, the orbit is not closed and has infinite length. As it remains confined to the surface of the torus, it is called *quasiperiodic*. Such bounded, yet infinitely long orbits, cannot be observed in 2D flows.

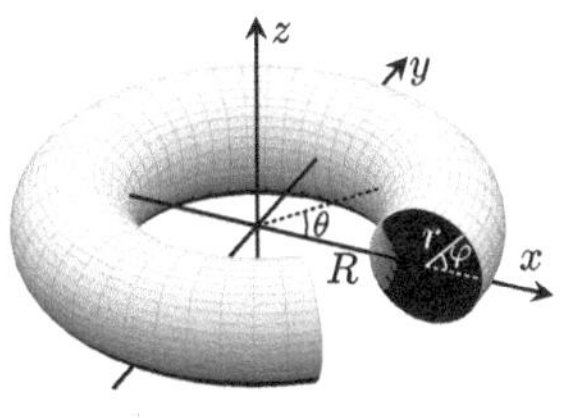

Figure 4.14: Illustration of a torus. Orbits confined to its surface can combine rotation around the axis of the torus and rotation around the axis of the tube, and be doubly periodic. The equations (4.34) describe the torus surface in terms of the angles ϕ and θ.

Thus, it appears that the Poincaré–Bendixson theorem is not applicable to 3D flows. Consequently, chaotic dynamics can be observed in 3D flows: Although phase trajectories cannot cross each other, they can bend around by exploiting the third dimension. Next, we consider two examples of 3D flows in which chaotic dynamics is observed.

Example 4.12. The **Lorenz system** was designed to model convective flows of air in a circular pipe erected vertically and heated at the bottom. The model is represented by the following equations:

$$\begin{cases} \dot{x} = \sigma(y - x), \\ \dot{y} = x(\rho - z) - y, \\ \dot{z} = xy - \beta z, \end{cases} \tag{4.35}$$

where the variables x, y, and z represent the flow rate, the top-to-bottom temperature difference, and the deviation of temperature from the linear profile, respectively. The model has three (positive) parameters: σ—at the Prandtl number, ρ—the Rayleigh number, and β—a geometric factor describing the pipe. This model is famous for exhibiting chaotic dynamics and highlighting the inherent difficulties of weather forecasting. Let us analyse it.

Equilibria:
Equilibria are given by the roots of the following algebraic system:

$$\begin{cases} \sigma(y - x) = 0, \\ x(\rho - z) - y = 0, \\ xy - \beta z = 0. \end{cases}$$

From the first equation we obtain $x = y$. Substituting this condition into the third equation gives $z = x^2/\beta$. Finally, the second equation transforms into

$$x\left(\rho - \frac{x^2}{\beta} - 1\right) = 0.$$

This equation has three roots: $x_1 = 0$ and $x_{2,3} = \pm\sqrt{\beta(\rho - 1)}$. Thus, we have three equilibrium points

$$P_1 = (0, 0, 0), \quad P_{2,3} = (\pm\sqrt{\beta(\rho - 1)}, \pm\sqrt{\beta(\rho - 1)}, \rho - 1).$$

The last two equilibria exist only when $\rho > 1$. At $\rho = 1$, all three equilibria coincide, indicating that at this parameter value a *pitchfork bifurcation* occurs.

Stability:
To analyze the stability of equilibria, we calculate the Jacobian matrix

$$J = \begin{pmatrix} -\sigma & \sigma & 0 \\ \rho - z & -1 & -x \\ y & x & -\beta \end{pmatrix}.$$

For the equilibrium point $P_1 = (0,0,0)$, the Jacobian becomes

$$J(0,0,0) = \begin{pmatrix} -\sigma & \sigma & 0 \\ \rho & -1 & 0 \\ 0 & 0 & -\beta \end{pmatrix}.$$

The characteristic polynomial of this matrix is

$$\det(J(0,0,0) - \lambda I) = (-\beta - \lambda)[(\lambda + \sigma)(\lambda + 1) - \rho\sigma].$$

Thus, the eigenvalues are

$$\lambda_1 = -\beta, \quad \lambda_{2,3} = \frac{-(\sigma+1) \pm \sqrt{(\sigma-1)^2 + 4\sigma\rho}}{2}.$$

Since $\beta > 0$, we always have $\lambda_1 < 0$. The stability of P_1 therefore depends on $\lambda_{2,3}$: If $\rho < 1$, both $\lambda_{2,3} < 0$, so P_1 is a *stable node*; and if $\rho > 1$, one eigenvalue becomes positive, so P_1 is a *saddle point*.

At $\rho = 1$, we have a pitchfork bifurcation. Since $P_{2,3}$ exist for $\rho > 1$ and P_1 becomes unstable for $\rho > 1$, we conclude that at $\rho = 1$ a *supercritical pitchfork bifurcation* occurs, and the equilibria $P_{2,3}$ are stable. As eigenvalues change smoothly with variation of the parameter ρ, we deduce that for the equilibria $P_{2,3}$, λ_1 and λ_2 remain real and negative, while λ_3 decreases from zero and becomes negative.

The Jacobian at the equilibria

$$P_{2,3} = (\pm\sqrt{\beta(\rho-1)}, \pm\sqrt{\beta(\rho-1)}, \rho - 1)$$

is given by

$$J = \begin{pmatrix} -\sigma & \sigma & 0 \\ 1 & -1 & -\bar{x} \\ \bar{x} & \bar{x} & -\beta \end{pmatrix},$$

where $\bar{x} = \pm\sqrt{\beta(\rho-1)}$. The characteristic equations for the Jacobian in both equilibria are identical and given by

$$\lambda^3 + \lambda^2(\sigma + \beta + 1) + \lambda\beta(\sigma + \rho) + 2\sigma\beta(\rho - 1) = 0. \tag{4.36}$$

At the bifurcation point $\rho = 1$, the last term in (4.36) disappears and the eigenvalues are $\lambda_1 = -\beta, \lambda_2 = -(\sigma+1)$, and $\lambda_3 = 0$ (the same as for P_1). Eigenvalues cannot be found analytically for arbitrary values of ρ. However, it can be shown that an increase in ρ (from 1) results in an increase in λ_2 (from $-(\sigma+1)$) and a decrease in λ_3 (from 0) as is expected from our conclusion that at $\rho = 1$, we have a supercritical pitchfork bifurcation. With further increases in the bifurcation parameter ρ, eigenvalues λ_2 and λ_3 move towards

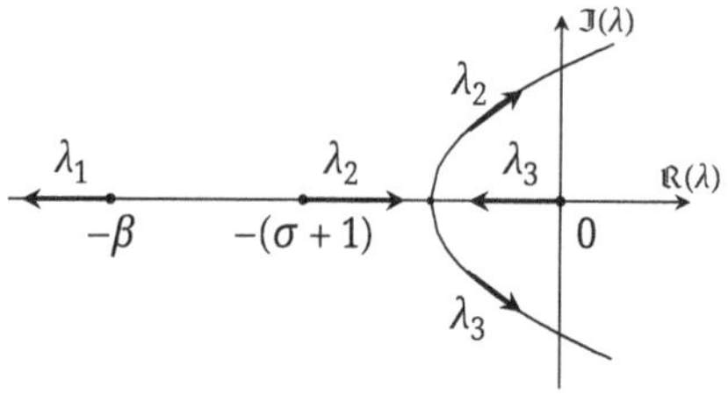

Figure 4.15: Evolution of eigenvalues for equilibria $P_{2,3}$ in the Lorenz system (4.35) as the parameter ρ increases from 1 to above ρ_{crit}.

each other and, at a certain value of ρ, become equal (illustrated in Figure 4.15). With further increases in ρ, these eigenvalues form a complex conjugate pair with negative real part, and the equilibria $P_{2,3}$ transform from stable nodes to stable node-spirals.

As ρ increases further, the real part of $\lambda_{2,3}$ increases and, at some critical value $\rho = \rho_{\text{crit}}$ becomes zero. The value of ρ_{crit} can be determined as follows. When $\rho = \rho_{\text{crit}}$, we have one real eigenvalue, $\lambda_1 = \mu$, and two imaginary eigenvalues, $\lambda_{2,3} = \pm i\nu$. The cubic in this case factorizes as

$$(\lambda+\mu)(\lambda^2+\nu^2) = \lambda^3 + \mu\lambda^2 + \nu^2\lambda + \mu\nu^2 = 0.$$

This is a factorization of the characteristic Eq. (4.36) when $\rho = \rho_{\text{crit}}$. Thus,

$$\mu = \sigma + \beta + 1, \quad \nu^2 = \beta(\sigma + \rho_{\text{crit}}), \quad \mu\nu^2 = 2\sigma\beta(\rho_{\text{crit}} - 1).$$

This gives the following equation for ρ_{crit}:

$$\mu\nu^2 = (\sigma+\beta+1)\beta(\sigma+\rho_{\text{crit}}) = 2\sigma\beta(\rho_{\text{crit}} - 1),$$

from which we obtain the required formula for ρ_{crit}

$$\rho_{\text{crit}} = \frac{\sigma(\sigma+\beta+3)}{\sigma-\beta-1}.$$

At this value of ρ, the Eq. (4.36) transforms into

$$(\lambda+\sigma+\beta+1)\left(\lambda^2 + \frac{2\beta\sigma(\sigma+1)}{\sigma-\beta-1}\right) = 0,$$

with eigenvalues

$$\lambda_1 = -(\sigma+\beta+1), \quad \lambda_{2,3} = \pm i\sqrt{\frac{2\beta\sigma(\sigma+1)}{\sigma-\beta-1}}.$$

At $\rho = \rho_{\text{crit}}$, the equilibria $P_{2,3}$ transform from stable node-spirals to saddle-spirals, so that a *Hopf bifurcation* takes place.

This bifurcation is supercritical, and this can be shown in the following way. Let us consider an arbitrary orbit $\mathbf{r}(t) = (x(t), y(t), z(t))$ and check how the distance from this orbit to the point $\mathbf{r}_0 = (0, 0, \rho + \sigma)$ changes over time as follows:

$$\begin{aligned}\frac{1}{2}\frac{d}{dt}(\mathbf{r}(t) - \mathbf{r}_0)^2 = \frac{1}{2}\frac{d}{dt}(x^2 + y^2 + (z - \rho - \sigma)^2) &= x\dot{x} + y\dot{y} + \dot{z}(z - \rho - \sigma)\\ &= \sigma x(y - x) + y(\rho x - y - xz) + (-\beta z + xy)(z - \rho - \sigma)\\ &= \cdots = -\sigma x^2 - y^2 - \beta\left[z - \frac{1}{2}(\rho + \sigma)\right]^2 + \frac{1}{4}\beta(\rho + \sigma)^2.\end{aligned}$$

This shows that $|\mathbf{r}(t) - \mathbf{r}_0|$ decreases outside the ellipsoid

$$\frac{x^2}{\frac{\beta}{4\sigma}(\rho + \sigma)^2} + \frac{y^2}{\frac{\beta}{4}(\rho + \sigma)^2} + \frac{[z - \frac{1}{2}(\rho + \sigma)]^2}{\frac{1}{4}(\rho + \sigma)^2} = 1.$$

This implies the existence of an attractor inside the above ellipsoid for any value of ρ. However, when $\rho > \rho_{\text{crit}}$, there are no stable equilibria in the system, and from this we conclude that the transition of $P_{2,3}$ into saddle-spirals should be accompanied with formation of stable limit circles.

Formation of chaos:
In some range of values of ρ, two limit cycles act as separate attractors. As the value of ρ increases, the sizes of these limit cycles grow, and eventually they begin to interfere with each other: Their domains of attraction overlap, and the two stable attractors merge into a single one, called the *Lorenz attractor*, which has a so-called *fractal structure.* At this stage, the system exhibits chaotic dynamics, as illustrated in Figure 4.16. Attractors associated with chaotic dynamics are called *strange attractors*.

Example 4.13. Rössler system is a three-dimensional flow represented by the following equations:

$$\begin{cases}\dot{x} = -y - z,\\ \dot{y} = x + ay,\\ \dot{z} = b + z(x - c),\end{cases} \tag{4.37}$$

where a, b, and c are positive parameters.

This system also exhibits chaotic behavior, but the route to chaos in the Rössler system is different from that in the Lorenz system. It is more closely related to the *period-doubling cascade* observed in the logistic map analyzed in Chapter 2.

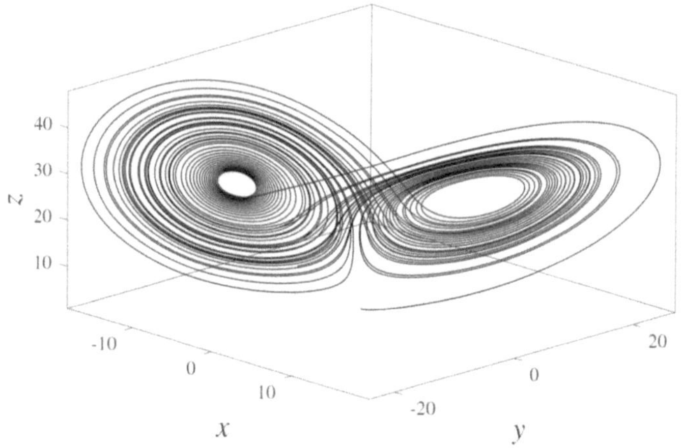

Figure 4.16: Phase trajectory exhibiting chaotic dynamics in the Lorenz system (4.35) with parameters $\sigma = 10$, $\rho = 28$, and $\beta = 8/3$. Starting point $(x, y, z) = (0, 1, 1.05)$.

Equilibria:

The equilibria of the Rössler system are given by the roots of the algebraic equations

$$\begin{cases} -y - z = 0, \\ x + ay = 0, \\ b + z(x - c) = 0. \end{cases}$$

From the first equation we obtain $z = -y$, and from the second $x = -ay$. Substituting both into the third equation yields

$$b - y(-ay - c) = 0 \quad \Rightarrow \quad ay^2 + cy - b = 0,$$

with solutions

$$y_{1,2} = \frac{-c \pm \sqrt{c^2 + 4ab}}{2a}.$$

Correspondingly,

$$x_{1,2} = -ay_{1,2}, \quad z_{1,2} = -y_{1,2}.$$

Hence, the system has two equilibria, i. e.,

$$P_{\pm} = \left(\frac{c \pm \sqrt{c^2 - 4ab}}{2}, -\frac{c \pm \sqrt{c^2 - 4ab}}{2a}, \frac{c \pm \sqrt{c^2 - 4ab}}{2a} \right),$$

which exist provided that $c^2 > 4ab$. Therefore, we conclude that at $c^2 = 4ab$ the system undergoes a *saddle-node bifurcation* at point

$$P = \left(\frac{c}{2}, -\frac{c}{2a}, \frac{c}{2a}\right).$$

Stability:
The Jacobian of the system is

$$J = \begin{pmatrix} 0 & -1 & -1 \\ 1 & a & 0 \\ z & 0 & x-c \end{pmatrix},$$

with the characteristic equation

$$-\lambda^3 + (a + x - c)\lambda^2 + (ac - ax - 1 - z)\lambda + x - c + az = 0.$$

At the bifurcation point, we have $x = c/2$ and $z = c/2a$, and thus the characteristic equation becomes

$$[\lambda^2 - (a - c/2)\lambda + (1 + c/2a - ac/2)]\lambda = 0.$$

At the bifurcation point, two eigenvalues of the Jacobian, $\lambda_{1,2}$, are complex conjugates with a positive real part and the real eigenvalue, $\lambda_3 = 0$. With an increase of c, the equilibrium P_- becomes a *saddle-focus*, which is a combination of an unstable focus with a stable manifold ($\lambda_3 < 0$). Phase trajectories near this equilibrium converge towards it approximately along the z-direction and spiral out in the x–y plane.

At the same time, P_+ becomes an *unstable node-focus*. With a further increase of the parameter c, it undergoes a (degenerate) Hopf bifurcation and transforms into a *saddle-focus*, namely, a stable focus combined with an unstable manifold. Phase trajectories in the vicinity of this equilibrium spiral in within a plane defined by the spiral eigenvectors and then leave along the direction of the unstable manifold.

The linkage between these two local behaviors results in periodic motion along the limit cycle shown in Figure 4.17A. A further increase in the parameter c leads to a period-doubling cascade, which is very similar to that observed in the logistic map. This cascade culminates in a chaotic solution (see Figure 4.17C), which is observed for sufficiently large values of the parameter c.

As we see, the chaotic regime in a 3D flow is associated with the existence of a strange attractor, which can be viewed as a set of irregular and aperiodic stable limit cycles. As the convergence to an attractor is associated with dissipative regimes, we conclude that chaos is observed in 3D flows forming dissipative systems. Whether the 3D flow represents an area-expanding, dissipative, or conservative system depends on the trace of its Jacobian

$$\operatorname{tr} J(x,y,z) = \frac{\partial f(x,y,z)}{\partial x} + \frac{\partial g(x,y,z)}{\partial y} + \frac{\partial h(x,y,z)}{\partial z},$$

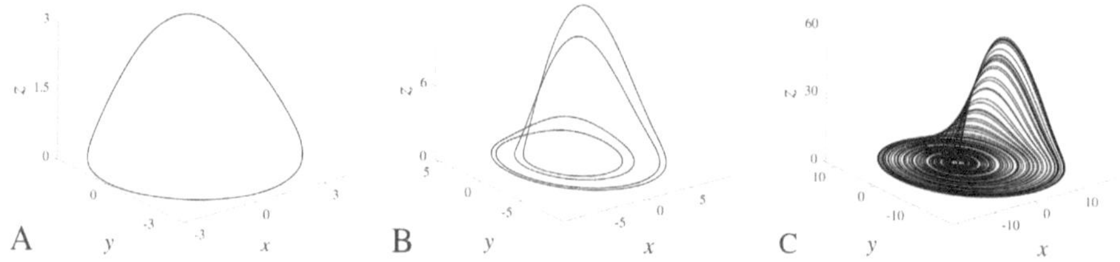

Figure 4.17: Evolution of the limit cycle in the Rössler system (4.37) with variation of the bifurcation parameter c. **Panel A:** Ordinary limit cycle when $c = 2$. **Panel B:** Two-periodic limit cycle when $c = 4$. **Panel C:** Irregular limit cycle (chaotic regime) when $c = 9$. Other parameters: $a = 0.2$, $b = 0.2$.

in the same way as in 2D flows. Particularly, we can check that for the Lorenz system

$$\operatorname{tr} J(x, y, z) = -\sigma - 1 - \beta < 0$$

since σ and β are both positive, and therefore this system is dissipative.

4.14 Summary

In this chapter, we have considered multidimensional flows. We have found that, while the solutions to 1D flows (discussed in the previous chapter) can only be monotonic, solutions to 2D and 3D flows are far more complicated. Firstly, this concerns the dynamics of flows in the vicinity of equilibria, as reflected by the variety of equilibrium types we have identified. This variety includes nodes, saddles, spirals, and centers in 2D flows, and their further combinations in 3D flows.

We have extended qualitative methods introduced for the study of 1D flows to analyze solutions to 2D flows, as represented by the orbits on the phase plane. Tools used for such analysis include nullclines, which divide the phase plane into domains with different directions of flow, and stable and unstable manifolds of saddles. The latter serve as heteroclinic orbits and separatrices and shed light on the global pattern formed by the flows.

Furthermore, we have identified periodic solutions, called *limit cycles*, which occur in two-dimensional flows but cannot arise in one-dimensional systems. From this we may conclude that the dynamical regimes exhibited by nonlinear planar flows include:
- Multistability, discussed in the previous chapter and illustrated by the Lotka–Volterra competition model (Example 4.9); and
- Oscillations, exemplified by the van der Pol model (Example 4.3).

Another nonlinear phenomenon characteristic of two-dimensional flows is *excitability*. Although it plays an important role in biological applications [17], it is often omitted from standard dynamical-systems textbooks. We likewise do not pursue excitable dynamics here.

The types of dynamical regimes that can be observed in 2D flows are restricted by the Poincaré–Bendixson theorem, which excludes the formation of chaos in two-dimensional systems. We have seen that chaotic regimes can arise in 3D flows, and these regimes are associated with the existence of stable limit circles that are aperiodic and shaped as infinitely long lines.

We have also considered multivariable flows with parameters. We have found that bifurcation occurs when one of the eigenvalues of the Jacobian at an equilibrium becomes zero. In this case, saddle-node, transcritical, or pitchfork bifurcations are observed. These types of bifurcations are already familiar from the study of 1D flows. In addition to these, there exists the so-called *Andronov–Hopf* bifurcation, which is associated with the formation of stable and unstable limit cycles in multidimensional flows. This bifurcation corresponds to transitions between stable and unstable spirals, occurring when the real part of a pair of complex eigenvalues becomes zero.

It appears that the level of complexity of 1D maps, as reflected by the dynamical regimes observed, is only reached in 3D flows, where period-doubling cascades and chaos can also be found.

Further reading on the theory of multidimensional flows can be found in a number of textbooks including [11, 12, 13], while their applications to biological systems are illustrated in [4, 17].

4.15 Exercises

Problem 4.1. Consider the dynamical system

$$\begin{cases} \dot{x} = 2x(1-y), \\ \dot{y} = 2 - y - x^2. \end{cases} \tag{4.38}$$

(i) Sketch the vector field of this system using the nullclines.
(ii) Find all equilibria and determine their types.
(iii) Draw the qualitative phase portrait of the system.

Problem 4.2. Consider the dynamical system

$$\begin{cases} \frac{dx}{dt} = by(1-y) - y^2 x, \\ \frac{dy}{dt} = (1-y)(1-by) + y^2 x, \end{cases}$$

where $b > 0$ is a parameter. Find the equilibrium point of this system and describe how the type of this point depends on the value of parameter b.

Problem 4.3 (The Kermack–McKendrick model). The transmission of a disease is described by the model represented by the system of differential equations

$$\begin{cases} \dot{S} = -aSI, \\ \dot{I} = aSI - bI, \\ \dot{R} = bI, \end{cases} \tag{4.39}$$

where a and b are positive constants and $S(t), I(t)$, and $R(t)$ are the numbers of susceptible, infected, and recovered individuals. Show that:

(i) The overall population $N = S + I + R$ remains constant. Then consider the dynamics in the (S, I) projected phase plane and sketch the corresponding phase portrait.

(ii) A trajectory with initial values $S(0) = S_0$ and $I(0) = I_0$ satisfies

$$I = I_0 + S_0 - S + \frac{b}{a} \ln\left(\frac{S}{S_0}\right). \tag{4.40}$$

(iii) If $S_0 < b/a$, then the disease dies out monotonically, that is, I monotonically declines to zero.

(iv) If $S_0 = b/a + \delta$ with $0 < \delta \ll 1$, then there is an epidemic, which peaks and then dies out in such a way that $S \to b/a - \sqrt{2I_0 b/a + \delta^2}$, provided that $I_0 \ll 1$.

4.16 Solutions to exercises

Solution to Problem 4.1.

(i) *Nullclines and vector field sketch.* The x-nullclines are given by $\dot{x} = 0$, i. e.,

$$x = 0 \quad \text{or} \quad y = 1.$$

The y-nullcline is given by $\dot{y} = 0$, i. e.,

$$y = 2 - x^2,$$

a downward-opening parabola with vertex at $(0, 2)$.

Direction of flow:

- $\dot{x} > 0$ when either $x > 0$ and $y < 1$, or $x < 0$ and $y > 1$, and $\dot{x} < 0$ in the rest of the plane.
- $\dot{y} > 0$ below the parabola $y = 2 - x^2$, and negative above it.

Flow directions in all domains separated by the nullclines are shown in Figure 4.18A.

(ii) *Equilibria and their types.* Equilibria occur at intersections of the nullclines that are in three points $(X_1, Y_1) = (0, 2)$, $(X_2, Y_2) = (-1, 1)$ and $(X_3, Y_3) = (1, 1)$.

Stability of equilibrium points is defined by eigenvalues of the Jacobian

$$J(x, y) = \begin{pmatrix} \partial_x \dot{x} & \partial_y \dot{x} \\ \partial_x \dot{y} & \partial_y \dot{y} \end{pmatrix} = \begin{pmatrix} 2(1 - y) & -2x \\ -2x & -1 \end{pmatrix}.$$

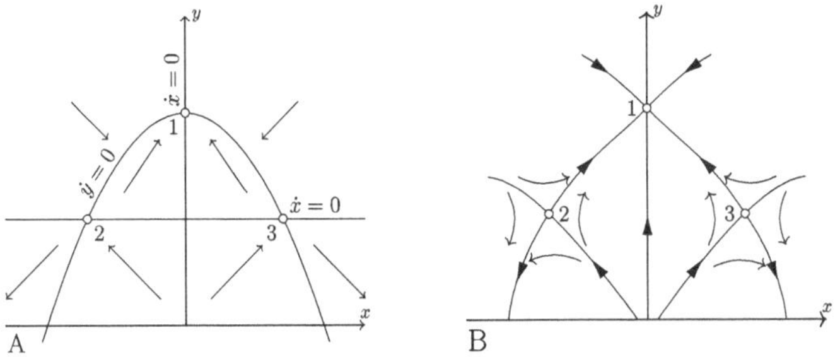

Figure 4.18: Panel A: Nullclines and direction field for the system (4.38). **Panel B:** Sketch of flows for this system. Equilibria are indicated and numbered as follows: 1—stable node, 2 and 3—saddles.

– At $(0,2)$:

$$J(0,2) = \begin{pmatrix} -2 & 0 \\ 0 & -1 \end{pmatrix}, \quad \lambda_1 = -2,\ \lambda_2 = -1.$$

Both eigenvalues are negative, and therefore the fixed point is a *stable node.*

– At $(1,1)$:

$$J(1,1) = \begin{pmatrix} 0 & -2 \\ -2 & -1 \end{pmatrix}, \quad \text{tr} = -1,\ \det = -4 < 0.$$

The determinant is negative, and therefore the fixed point is a *saddle.*

– At $(-1,1)$:

$$J(-1,1) = \begin{pmatrix} 0 & 2 \\ 2 & -1 \end{pmatrix}, \quad \text{tr} = -1,\ \det = -4 < 0.$$

The determinant is negative, and therefore the fixed point is a *saddle.*

(iii) *Qualitative phase portrait.* The stable node at $(0,2)$ lies on the unstable manifolds of both saddles. The saddles at $(\pm 1,1)$ possess separatrices (stable manifolds) that partition the plane into regions, where trajectories either converge to the stable node or diverge to infinity. A global picture, illustrating trajectories attracted to the stable node with saddle separatrices organising the phase space, is sketched in Figure 4.18B.

Solution to Problem 4.2. The equilibrium points are found as solutions of the algebraic system

$$\begin{cases} by(1-y) - y^2x = 0, \\ (1-y)(1-by) + y^2x = 0. \end{cases}$$

From the first equation we obtain $y = 0$ or $x = \frac{b(1-y)}{y}$. The case $y = 0$ does not satisfy the second equation

$$(1-y)(1-by) + y^2x = 0 \quad \Rightarrow \quad 1 \neq 0,$$

and is therefore excluded.

After substituting $x = \frac{b(1-y)}{y}$ into the second equation we obtain

$$(1-y)(1-by) + y^2 \cdot \frac{b(1-y)}{y} = 0 \quad \Rightarrow \quad y = 1.$$

Hence, the unique equilibrium point of the system is

$$(X, Y) = (0, 1).$$

The Jacobian at this point is

$$J(0,1) = \begin{pmatrix} -1 & -b \\ 1 & b-1 \end{pmatrix}, \quad \operatorname{tr} J = b-2, \quad \det J = 1.$$

The characteristic polynomial is

$$\lambda^2 - (b-2)\lambda + 1 = 0,$$

with discriminant

$$D = (b-2)^2 - 4 = b(b-4).$$

Classification of the equilibrium:

$0 < b < 2$	Stable spiral (complex eigenvalues with negative real part).
$b = 2$	Center (purely imaginary eigenvalues; Hopf bifurcation).
$2 < b < 4$	Unstable spiral (complex eigenvalues with positive real part).
$b = 4$	Degenerate unstable node (repeated positive eigenvalue).
$b > 4$	Unstable node (two distinct positive real eigenvalues).

We observe that at $b = 2$ there is a transition from a stable to an unstable focus, suggesting the occurrence of a non-degenerate Hopf bifurcation. Consequently, for $b > 2$ the equilibrium becomes an unstable spiral, and a stable limit cycle emerges. This can be rigorously established by computing the first Lyapunov coefficient at the equilibrium when $b = 2$.

Solution to Problem 4.3.

(i) Let $N(t) = S(t) + I(t) + R(t)$. Then,

$$\dot{N} = \dot{S} + \dot{I} + \dot{R} = -aSI + (aSI - bI) + bI = 0,$$

so N remains constant.

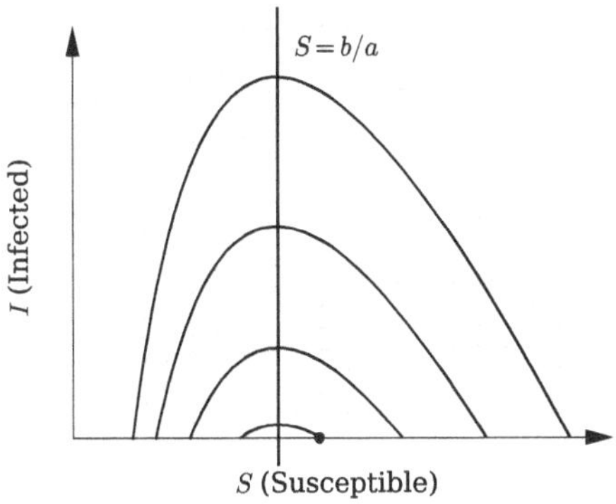

Figure 4.19: Nullcline $\dot{I} = 0$, direction field and sample orbits for the Kermack–McKendrick model (4.39). Small dot marks the initial point of the orbit considered in (d).

In the (S, I) phase plane (with $R = N - S - I$), the nullclines are

$$\dot{S} = 0 \quad \Rightarrow \quad I = 0, \quad \dot{I} = 0 \quad \Rightarrow \quad I = 0 \quad \text{or} \quad S = \frac{b}{a}.$$

It follows that all points on $I = 0$ are equilibria. $\dot{I} > 0$ if $S > b/a$ and negative otherwise.
To sketch the phase portrait we note that S and I are positive quantities, and therefore we need to consider the first quadrant only. Also, since $I + S \leq N$ phase trajectories lie beneath the line, $I + S = N$. The phase portraits shown in Figure 4.19. Phase trajectories are represented by concave dawn curves. Phase trajectories starting with $S > b/a$ have a maximum at the nullcline $S = \frac{b}{a}$.

(ii) Equation for phase trajectories are

$$\frac{dI}{dS} = \frac{\dot{I}}{\dot{S}} = \frac{aSI - bI}{-aSI} = -1 + \frac{b}{aS}.$$

Integrating gives

$$I + S - \frac{b}{a} \ln S = C.$$

Applying the initial conditions (S_0, I_0), we get

$$I = I_0 + S_0 - S + \frac{b}{a} \ln\left(\frac{S}{S_0}\right).$$

(iii) If $S_0 < \frac{b}{a}$, then $aS_0 - b < 0$. Since $S(t)$ decreases whenever $I(t) > 0$, we have $S(t) \leq S_0$ for all $t \geq 0$, hence $aS(t) - b \leq aS_0 - b < 0$. Therefore,

$$\dot{I} = I(aS - b) < 0,$$

so $I(t)$ decreases monotonically to 0.

(iv) Let $S_0 = \frac{b}{a} + \delta$ with $0 < \delta \ll 1$, $I_0 \ll 1$. This initial conditions is marked by the black dot in Figure 4.19.
Set $s = \frac{b}{a}$, and let the final susceptible level be $S_\infty = s - x$ (with $x > 0$). Using the Eq. (4.40) with $I = 0$ as $t \to \infty$, we have

$$0 = I_0 + S_0 - S_\infty + s\ln\left(\frac{S_\infty}{S_0}\right) = I_0 + \delta + x + s\left[\ln\left(1 - \frac{x}{s}\right) - \ln\left(1 + \frac{\delta}{s}\right)\right].$$

Using the quadratic approximations $\ln(1+y) \approx y - \frac{y^2}{2}$ (valid for small y), we transform the last term

$$s\left[\ln\left(1 - \frac{x}{s}\right) - \ln\left(1 + \frac{\delta}{s}\right)\right] \approx -x - \delta - \frac{x^2 - \delta^2}{2s}.$$

Hence,

$$0 \approx I_0 + \delta + x - \left(x + \delta + \frac{x^2 - \delta^2}{2s}\right) = I_0 - \frac{x^2 - \delta^2}{2s}.$$

From here,

$$x \approx \sqrt{\delta^2 + \frac{2b}{a} I_0},$$

and, therefore,

$$S_\infty = s - x = \frac{b}{a} - \sqrt{\delta^2 + \frac{2b}{a} I_0}.$$

Thus,

$$S \to \frac{b}{a} - \sqrt{2I_0 \frac{b}{a} + \delta^2} \quad \text{as } t \to \infty.$$

Since at $t = 0$ we have $S_0 > b/a$, then according to the result obtained in (iii) the number of infected individuals, I, increases, leading to an epidemic. However, the number of susceptible individuals decreases over time, and once it drops below b/a the number of infected individuals, I, also begins to decrease and eventually declines to 0 when $S = S_\infty$.

5 Two-variable discrete-time dynamical systems

Two-dimensional maps are discrete-time dynamical systems whose state is described by two variables, say x and y. These maps are defined by two coupled difference equations

$$\begin{cases} x_{n+1} = f(x_n, y_n), \\ y_{n+1} = g(x_n, y_n), \end{cases} \tag{5.1}$$

where the functions $f(x,y)$ and $g(x,y)$ are usually assumed to have continuous first partial derivatives. Quite often, the variables in this system are interpreted as the coordinates of a point in the plane, so that the system defines a mapping of the plane.

Solutions of 2D maps are represented by consecutive sets of points on the plane (x_i, y_i), $i = 0, 1, \dots, n, \dots$ and called orbits. The orbit for the given map is uniquely defined by the initial point (x_0, y_0).

We have already encountered a 2D map in Chapter 2 when we considered the Fibonacci model given by the second-order difference equation

$$x_{n+1} = x_n + x_{n-1}.$$

By introducing an extra variable $y_n = x_{n-1}$, we transform this equation into

$$\begin{cases} x_{n+1} = x_n + y_n \\ y_{n+1} = x_n, \end{cases}$$

which is a 2D map.

The concepts of fixed points and their stability, introduced in the previous chapters, can be extended naturally to two-dimensional maps.

Definition 5.1. The state (x, y) of the system (5.1) is called **stationary** if, given $(x_0, y_0) = (x, y)$, it follows that $(x_n, y_n) = (x, y)$ for all n. This state is also known as a **fixed point** or **equilibrium point**.

It follows that the fixed point is a root of the algebraic system

$$\begin{cases} x = f(x, y), \\ y = g(x, y). \end{cases} \tag{5.2}$$

Definition 5.2. The fixed point (x, y) of the system (5.1) is **stable** if, for any ε, there exists δ such that, from $(x_0 - x)^2 + (y_0 - y)^2 < \delta$, it follows that $(x_n - x)^2 + (y_n - y)^2 < \varepsilon \; \forall n$.

Definition 5.3. The fixed point (x, y) of the system (5.1) is **asymptotically stable** if there exists δ such that, from $(x_0 - x)^2 + (y_0 - y)^2 < \delta$, it follows that $(x_n, y_n) \to (x, y)$ as $n \to \infty$.

https://doi.org/10.1515/9783111348339-005

5.1 Linear map

As with previous dynamical systems, we will start with a linear case. A linear and homogeneous 2D map is defined by the system

$$\begin{cases} x_{n+1} = a_{11}x_n + a_{12}y_n \\ y_{n+1} = a_{21}x_n + a_{22}y_n, \end{cases} \tag{5.3}$$

where the coefficients a_{ij} are constants. We can rewrite this system in a vector form

$$\mathbf{u}_{n+1} = A\mathbf{u}_n, \tag{5.4}$$

where $\mathbf{u}_n = \binom{x_n}{y_n}$ and $A = \begin{pmatrix} a_{11} & a_{12} \\ a_{21} & a_{22} \end{pmatrix}$.

Note that the fixed points of the system (5.3) are given by the roots of the equation $\mathbf{u} = A\mathbf{u}$, or equivalently $(A - I)\mathbf{u} = \mathbf{0}$. It follows that the point $(X, Y) = (0, 0)$ is the only fixed point provided that the matrix $A - I$ is non-singular. This condition implies that

$$\det(A - I) \neq 0 \quad \Rightarrow \quad \begin{vmatrix} a_{11} - 1 & a_{12} \\ a_{21} & a_{22} - 1 \end{vmatrix} \neq 0 \quad \Rightarrow \quad \det A - \operatorname{tr} A + 1 \neq 0,$$

so the condition for the existence of a single fixed point, expressed in terms of the properties of matrix A, can be written as

$$\operatorname{tr} A \neq \det A + 1. \tag{5.5}$$

Theorem 5.1. *Assume that the matrix A has two linearly independent eigenvectors. Then, the general solution of the system* (5.3) *is*

$$\mathbf{u}_n = c_1\mathbf{b}_1\lambda_1^n + c_2\mathbf{b}_2\lambda_2^n, \tag{5.6}$$

where $(\lambda_1, \mathbf{b}_1)$ *and* $(\lambda_2, \mathbf{b}_2)$ *are eigensolutions of the matrix A, and* c_1 *and* c_2 *are arbitrary constants.*

Proof. Since $(\lambda_1, \mathbf{b}_1)$ and $(\lambda_2, \mathbf{b}_2)$ are eigensolutions of the matrix A,

$$A\mathbf{b}_1 = \lambda_1\mathbf{b}_1,$$
$$A\mathbf{b}_2 = \lambda_2\mathbf{b}_2.$$

Let us introduce the matrix $B = (\mathbf{b}_1, \mathbf{b}_2)$. Note that this matrix is invertible since vectors $\mathbf{b}_1$ and $\mathbf{b}_2$ are linearly independent. Making the implicit substitution $\mathbf{u}_n = B\mathbf{v}_n$, we obtain

$$\mathbf{u}_{n+1} = A\mathbf{u}_n \quad \Rightarrow \quad B\mathbf{v}_{n+1} = AB\mathbf{v}_n.$$

From this equation, we need to work out the product AB as follows:

$$AB = A(\mathbf{b}_1, \mathbf{b}_2) = (A\mathbf{b}_1, A\mathbf{b}_2) = (\lambda_1 \mathbf{b}_1, \lambda_2 \mathbf{b}_2) = BD,$$

where $D = \begin{pmatrix} \lambda_1 & 0 \\ 0 & \lambda_2 \end{pmatrix}$ is a diagonal matrix. Thus, we have proved that

$$B\mathbf{v}_{n+1} = BD\mathbf{v}_n.$$

As we noted before, the matrix B is invertible, and therefore we conclude that

$$\mathbf{v}_{n+1} = D\mathbf{v}_n.$$

It follows that
For $n = 0$: $\mathbf{v_1} = D\mathbf{v_0}$.
For $n = 1$: $\mathbf{v_2} = D\mathbf{v_1} = D(D\mathbf{v_0}) = D^2\mathbf{v_0}$.
For $n = 2$: $\mathbf{v_3} = D\mathbf{v_2} = D^3\mathbf{v_0}$.
⋮
For any n: $\mathbf{v}_n = D^n\mathbf{v_0}$.

By introducing the components of $\mathbf{v}_0$,

$$\mathbf{v_0} = \begin{pmatrix} c_1 \\ c_2 \end{pmatrix},$$

we can rewrite the formula for $\mathbf{v}_n$ as

$$\mathbf{v}_n = \begin{pmatrix} \lambda_1^n & 0 \\ 0 & \lambda_2^n \end{pmatrix} \begin{pmatrix} c_1 \\ c_2 \end{pmatrix} = \begin{pmatrix} \lambda_1^n c_1 \\ \lambda_2^n c_2 \end{pmatrix}.$$

Now, in terms of the variable $\mathbf{u}$ (back substitution),

$$\mathbf{u}_n = B\mathbf{v_n} = (\mathbf{b}_1, \mathbf{b}_2) \begin{pmatrix} \lambda_1^n c_1 \\ \lambda_2^n c_2 \end{pmatrix} = c_1 \mathbf{b}_1 \lambda_1^n + c_2 \mathbf{b}_2 \lambda_2^n. \qquad \square$$

Thus, the system (5.3) has a fixed point at the origin $(0, 0)$, and its properties can be analyzed based on the solutions given by (5.6). Similarly to the approach taken with continuous systems, we will consider cases associated with real and complex eigenvalues in order to classify the fixed points.

5.1.1 Real eigenvalues

If $c_2 = 0$, then the initial point lies on the line through $(0, 0)$ and parallel to $\mathbf{b}_1$ (called an *eigenline*)

$$\mathbf{u}_0 = c_1 \mathbf{b}_1 \lambda_1^0 + c_2 \mathbf{b}_2 \lambda_2^0 = c_1 \mathbf{b}_1.$$

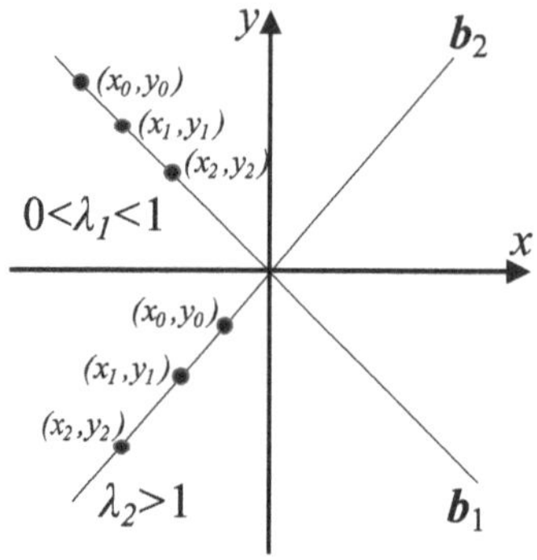

Figure 5.1: Fixed point at the origin with two illustrative orbits along the eigenlines $\mathbf{b}_1$ and $\mathbf{b}_2$. Each orbit is shown by three points representing the state of the system during the first three time steps. The eigenvalue λ_1 is assumed to lie between 0 and 1, and the points along the corresponding eigenline converge to the origin. The second eigenvalue, λ_2, is assumed to be greater than 1, and in this case the consecutive points of the orbit move away from the origin.

Furthermore, the entire orbit is located on this eigenline

$$\mathbf{u}_n = c_1\mathbf{b}_1\lambda_1^n + c_2\mathbf{b}_2\lambda_2^n = c_1\lambda_1^n\mathbf{b}_1 \parallel \mathbf{b}_1.$$

Moreover, if $|\lambda_1| < 1$, then $\mathbf{u}_n \to \mathbf{0}$ as $n \to \infty$, while if $|\lambda_1| > 1$, then $|\mathbf{u}_n| \to \infty$ as $n \to \infty$ (see Figure 5.1).

Similarly, if $c_1 = 0$, then the whole orbit lies on the eigenline parallel to $\mathbf{b}_2$ and converges to the origin if $|\lambda_2| < 1$, and diverges to infinity if $|\lambda_2| > 1$ (see Figure 5.1).

Definition 5.4. If $|\lambda_{1,2}| < 1$, then the fixed point $(0, 0)$ is asymptotically stable and is called a **stable node** or **attractor**. If either of the eigenvalues is negative, then it is also called a **flip attractor** since the orbit flips around the fixed point.

Definition 5.5. If $|\lambda_{1,2}| > 1$, then the fixed point is called an **unstable node** or **repeller** (**flip repeller** if at least one of the eigenvalues is negative).

Definition 5.6. If $|\lambda_1| < 1$ and $|\lambda_2| > 1$ (or vice versa), then the fixed point is unstable and is called a **saddle**.

If the matrix A has only one linearly independent eigenvector, then the formula (5.6) must be corrected, similarly to what was done for continuous systems in Chapter 4, so that the solution involves polynomial modulation.

Solution (5.6) also implies that if one of the eigenvalues is equal to 1, then there are infinitely many fixed points lying on the line parallel to the corresponding eigenvector, and none of them (including the origin) is asymptotically stable. This observation is consistent with condition (5.5), as it follows from the formula for the eigenvalues of matrix A

$$\lambda_{1,2} = \frac{\operatorname{tr} A \pm \sqrt{(\operatorname{tr} A)^2 - 4 \det A}}{2}.$$

By substituting $\operatorname{tr} A = \det A + 1$, we obtain

$$\lambda_{1,2} = \frac{\det A + 1 \pm \sqrt{(\det A + 1)^2 - 4\det A}}{2} = \frac{\det A + 1 \pm (\det A - 1)}{2}.$$

Thus,

$$\lambda_1 = \det A, \quad \lambda_2 = 1.$$

It follows that if $\det A = 1$ and $\operatorname{tr} A = 2$ (i. e., A is an identity matrix), then every point in the phase plane is an equilibrium point (provided that A has two linearly independent eigenvectors).

5.1.2 Complex eigenvalues

Although A is a matrix with real entries, it may still possess complex eigenvalues and complex eigenvectors. In this case the solution is complex, and we must extract its real part. Let us represent the eigensolutions of the matrix A in the form

$$\begin{cases} \lambda_{1,2} = \alpha \pm i\beta, \\ \mathbf{b}_{1,2} = \mathbf{b}_r \mp i\mathbf{b}_{im}, \end{cases} \tag{5.7}$$

where α and β are real numbers, and $\mathbf{b}_r$ and $\mathbf{b}_{im}$ are vectors with real components.

The solution (5.6) in this case is represented by a pair of complex-conjugate expressions, whose real part describes the actual trajectory in the plane. This property is formalized in the following theorem.

Theorem 5.2. *If the eigenvalues of the matrix A in* (5.4) *are complex, then the real part of the solution* (5.6), *written using the notations introduced in* (5.7), *is*

$$\mathbf{u}_n = |\lambda|^n[(c_1\cos(n\varphi) - c_2\sin(n\varphi))\mathbf{b}_r + (c_1\sin(n\varphi) + c_2\cos(n\varphi))\mathbf{b}_{im}],$$

where $|\lambda| = \sqrt{\alpha^2 + \beta^2}$, $\tan\varphi = \beta/\alpha$, and c_1, c_2 are arbitrary constants. This can also be written as

$$\mathbf{u}_n = (\mathbf{b}_r, \mathbf{b}_{im})|\lambda|^n \begin{pmatrix} \cos(n\varphi) & -\sin(n\varphi) \\ \sin(n\varphi) & \cos(n\varphi) \end{pmatrix} \begin{pmatrix} c_1 \\ c_2 \end{pmatrix}.$$

Proof. Let us introduce the matrix

$$B = (\mathbf{b}_r, \mathbf{b}_{im}),$$

and make the substitution

$$\mathbf{u}_n = B\mathbf{v}_n.$$

Then,

$$\mathbf{u}_{n+1} = A\mathbf{u}_n \quad \Rightarrow \quad \mathbf{u}_{n+1} = B\mathbf{v}_{n+1} = A\mathbf{u}_n = AB\mathbf{v}_n.$$

Now, let us compute the product AB. From the definition of an eigensolution, $A\mathbf{b}_1 = \lambda_1\mathbf{b}_1$, we obtain

$$\begin{aligned} A\mathbf{b}_1 &= A(\mathbf{b}_r + \imath\mathbf{b}_{im}) = A\mathbf{b}_r + iA\mathbf{b}_{im} = \lambda_1\mathbf{b}_1 \\ &= (\alpha + \imath\beta)(\mathbf{b}_r + \imath\mathbf{b}_{im}) \\ &= (\alpha\mathbf{b}_r - \beta\mathbf{b}_{im}) + i(\beta\mathbf{b}_r + \alpha\mathbf{b}_{im}). \end{aligned}$$

Equating real and imaginary parts in the above equation, we obtain

$$\begin{aligned} A\mathbf{b}_r &= \alpha\,\mathbf{b}_r + \beta\,\mathbf{b}_{im}, \\ A\mathbf{b}_{im} &= -\beta\,\mathbf{b}_r + \alpha\,\mathbf{b}_{im}. \end{aligned}$$

Now,

$$\begin{aligned} AB &= A(\mathbf{b}_r, \mathbf{b}_{im}) = (A\mathbf{b}_r, A\mathbf{b}_{im}) \\ &= (\alpha\mathbf{b}_r + \beta\mathbf{b}_{im}, -\beta\mathbf{b}_r + \alpha\mathbf{b}_{im}) = (\mathbf{b}_r, \mathbf{b}_{im})\begin{pmatrix} \alpha & -\beta \\ \beta & \alpha \end{pmatrix}. \end{aligned}$$

We can rewrite the matrix that appeared above in the following form:

$$\begin{pmatrix} \alpha & -\beta \\ \beta & \alpha \end{pmatrix} = \sqrt{\alpha^2 + \beta^2}\begin{pmatrix} \frac{\alpha}{\sqrt{\alpha^2+\beta^2}} & -\frac{\beta}{\sqrt{\alpha^2+\beta^2}} \\ \frac{\beta}{\sqrt{\alpha^2+\beta^2}} & \frac{\alpha}{\sqrt{\alpha^2+\beta^2}} \end{pmatrix}.$$

Introducing an angle φ such that

$$\cos\varphi = \frac{\alpha}{\sqrt{\alpha^2 + \beta^2}} = \frac{\alpha}{|\lambda|}, \quad \sin\varphi = \frac{\beta}{\sqrt{\alpha^2 + \beta^2}} = \frac{\beta}{|\lambda|},$$

we see that the matrix can be expressed as a scaling by $|\lambda| = \sqrt{\alpha^2 + \beta^2}$ (the modulus of the complex eigenvalues) followed by a rotation through angle φ (the argument of the complex eigenvalues), as expressed by the rotational matrix Y

$$Y = \begin{vmatrix} \cos(\varphi) & -\sin(\varphi) \\ \sin(\varphi) & \cos(\varphi) \end{vmatrix}.$$

Thus,

$$B\mathbf{v}_{n+1} = AB\mathbf{v}_n = B|\lambda|Y\mathbf{v}_n \quad \Rightarrow \quad \mathbf{v}_{n+1} = |\lambda|Y\mathbf{v}_n.$$

Altogether, the vector $\mathbf{v}_{n+1}$ is obtained from $\mathbf{v}_n$ by a counterclockwise rotation through angle φ, combined with a stretching or shrinking by the factor $|\lambda|$.

Let us compute $\mathbf{v}_n$

$$\begin{aligned}\mathbf{v_1} &= |\lambda| Y \mathbf{v_0},\\ \mathbf{v_2} &= |\lambda| Y \mathbf{v_1} = |\lambda|^2 Y^2 \mathbf{v_0}, \dots\\ \mathbf{v}_n &= |\lambda|^n Y^n \mathbf{v_0}.\end{aligned}$$

The matrix Y^2 is defined as

$$Y^2 = \begin{pmatrix} \cos(\varphi) & -\sin(\varphi) \\ \sin(\varphi) & \cos(\varphi) \end{pmatrix} \cdot \begin{pmatrix} \cos(\varphi) & -\sin(\varphi) \\ \sin(\varphi) & \cos(\varphi) \end{pmatrix} = \begin{pmatrix} \cos(2\varphi) & -\sin(2\varphi) \\ \sin(2\varphi) & \cos(2\varphi) \end{pmatrix}.$$

Consequently,

$$Y^n = \begin{pmatrix} \cos(n\varphi) & -\sin(n\varphi) \\ \sin(n\varphi) & \cos(n\varphi) \end{pmatrix},$$

and thus,

$$\mathbf{v}_n = |\lambda|^n Y^n \mathbf{v_0} = |\lambda|^n \begin{pmatrix} \cos(n\varphi) & -\sin(n\varphi) \\ \sin(n\varphi) & \cos(n\varphi) \end{pmatrix} \mathbf{v_0}.$$

Finally,

$$\mathbf{u}_n = B\mathbf{v}_n = (\mathbf{b}_r, \mathbf{b}_{im}) |\lambda|^n \begin{pmatrix} \cos(n\varphi) & -\sin(n\varphi) \\ \sin(n\varphi) & \cos(n\varphi) \end{pmatrix} \mathbf{v_0},$$

where

$$\mathbf{v_0} = \begin{pmatrix} c_1 \\ c_2 \end{pmatrix}.$$

□

The formula for $\mathbf{u}_n$ implies that each iteration corresponds to a counterclockwise rotation of the vector by the angle φ, together with a rescaling by the factor $|\lambda|$. It follows that the orbits of a linear map with complex eigenvalues are represented by a set of points lying on a logarithmic spiral centred at the origin of the coordinate system. This is reflected in the name of the fixed point $(0, 0)$ in this case, which is called a *focus* or *spiral*.

Definition 5.7. If the eigenvalues of the matrix A in system (5.4) are complex conjugates with modulus $|\lambda| < 1$, then the fixed point at the origin is called a **stable focus**.

Definition 5.8. If the eigenvalues of the matrix A in system (5.4) are complex conjugates with modulus $|\lambda| > 1$, then the fixed point at the origin is called an **unstable focus**.

Definition 5.9. If the eigenvalues of the matrix A in system (5.4) are complex conjugates with modulus $|\lambda| = 1$, then the fixed point at the origin is neutrally stable and is called a **centre**.

In the case when $|\lambda| = 1$, the orbit lies on an ellipse (or a circle), whose size is determined by the initial point of the orbit.

5.2 Nonlinear maps

Consider the general case of a nonlinear two-dimensional map given by Eq. (5.1). Fixed points of this map are obtained as solutions of the algebraic system (5.2). While in the case of a linear system the origin $(x, y) = (0, 0)$ is typically the only fixed point, in the nonlinear case there may exist multiple fixed points, not necessarily located at the origin.

Assume that (X, Y) is a fixed point of the nonlinear map and consider the orbit

$$\begin{pmatrix} x_0 \\ y_0 \end{pmatrix} \rightarrow \begin{pmatrix} x_1 \\ y_1 \end{pmatrix} \rightarrow \cdots \rightarrow \begin{pmatrix} x_n \\ y_n \end{pmatrix} \rightarrow \cdots$$

starting from an initial condition (x_0, y_0) in the neighborhood of (X, Y).

Introducing the deviation variables

$$\varepsilon_n = x_n - X, \quad \delta_n = y_n - Y,$$

the orbit can be described in terms of the sequence

$$\begin{pmatrix} \varepsilon_0 \\ \delta_0 \end{pmatrix}, \begin{pmatrix} \varepsilon_1 \\ \delta_1 \end{pmatrix}, \ldots, \begin{pmatrix} \varepsilon_n \\ \delta_n \end{pmatrix}, \ldots,$$

which represents the differences between the iterates (x_n, y_n) and the fixed point (X, Y). Convergence of the orbit (x_n, y_n) to the fixed point (X, Y) is therefore equivalent to

$$\varepsilon_n \rightarrow 0, \quad \delta_n \rightarrow 0 \quad \text{as } n \rightarrow \infty.$$

Taking into account that $x_n = X + \varepsilon_n$ and $y_n = Y + \delta_n$, we can rewrite the system (5.1) in terms of the new variables as follows:

$$\begin{aligned} x_{n+1} &= f(x_n, y_n) \quad \Rightarrow \quad X + \varepsilon_{n+1} = f(X + \varepsilon_n,\ Y + \delta_n), \\ y_{n+1} &= g(x_n, y_n) \quad \Rightarrow \quad Y + \delta_{n+1} = g(X + \varepsilon_n,\ Y + \delta_n). \end{aligned}$$

Assume that $\varepsilon_n \ll 1$ and $\delta_n \ll 1$, and apply the linear approximation

$$\begin{aligned} X + \varepsilon_{n+1} &\approx f(X, Y) + f_x(X, Y)\, \varepsilon_n + f_y(X, Y)\, \delta_n, \\ Y + \delta_{n+1} &\approx g(X, Y) + g_x(X, Y)\, \varepsilon_n + g_y(X, Y)\, \delta_n, \end{aligned}$$

where f_x and f_y denote the partial derivatives of f with respect to x and y, and g_x, g_y denote the partial derivatives of g with respect to x and y, all evaluated at (X, Y).

Since (X, Y) is a fixed point, we have $X = f(X, Y)$ and $Y = g(X, Y)$. Therefore, the deviations ε_{n+1} and δ_{n+1} satisfy

$$\varepsilon_{n+1} = f_x(X, Y)\, \varepsilon_n + f_y(X, Y)\, \delta_n,$$

$$\delta_{n+1} = g_x(X,Y)\,\varepsilon_n + g_y(X,Y)\,\delta_n.$$

This is the linear map, which can be written as

$$\begin{pmatrix}\varepsilon_{n+1}\\ \delta_{n+1}\end{pmatrix} = J\begin{pmatrix}\varepsilon_n\\ \delta_n\end{pmatrix},$$

where the matrix

$$J = \begin{pmatrix} f_x(X,Y) & f_y(X,Y)\\ g_x(X,Y) & g_y(X,Y)\end{pmatrix} \tag{5.8}$$

is the Jacobian of the map at the fixed point (X,Y). Based on the results obtained for linear maps, we conclude that

$$\begin{pmatrix}\varepsilon_n\\ \delta_n\end{pmatrix} \longrightarrow \begin{pmatrix}0\\ 0\end{pmatrix} \quad \text{if} \quad |\lambda_1| < 1 \text{ and } |\lambda_2| < 1,$$

where λ_1 and λ_2 are the eigenvalues of the Jacobian matrix J.

The classification of fixed points in nonlinear systems is based on the eigenvalues of the Jacobian at that point, in the same way as it is defined by the eigenvalues of the map-generating matrix in linear maps. If the eigenvalues are real and both lie in the interval $(-1,1)$, then the fixed point is called a *stable node* (or attractor). If the absolute values of both eigenvalues are greater than 1, then the fixed point is an *unstable node* (or repeller). If the absolute value of one eigenvalue is less than 1 and the other is greater than 1, the fixed point is a *saddle*. If the eigenvalues are complex conjugates with modulus less than 1, then the fixed point is a *stable focus*, and if the modulus is greater than 1, then the fixed point is an *unstable focus*.

The Jacobian at the fixed point gives a linear approximation of the orbits in the neighborhood of that point. Analysis of the global dynamics is based on the relative locations of fixed points and on the properties of stable and unstable manifolds of saddle points, in the same way as for nonlinear flows. Stable and unstable manifolds of a saddle point are defined similarly to how it was done for two-dimensional flows. There are orbits converging to a saddle, and the collection of all points belonging to these orbits forms a *stable manifold*. If we invert the map (so that the orbits do not change but the direction of motion along them is reversed), then the collection of all points belonging to those orbits which converge to the saddle forms an *unstable manifold*. These manifolds have the same properties as in flows. Stable manifolds act as repellers (orbits are directed away from stable manifolds), while unstable manifolds act as attractors (orbits tend to come closer to them). If there is a stable fixed point and a saddle point in the system, the stable point will always be located on the unstable manifold of the saddle point. A repeller (i. e., an unstable node) is always located on the stable manifold of the saddle. Stable manifolds are borders between basins of attraction of different attractors and are also called *separatrices*. A manifold connecting the saddle with another fixed point is called a *heteroclinic orbit*. The orbit connecting the saddle with itself (acting as both stable and unstable manifold) is called a *homoclinic orbit*.

Example 5.1. For the nonlinear map

$$\begin{cases} x_{n+1} = \frac{x_n}{2} + \frac{x_n^2}{4} + \frac{x_n y_n}{8}, \\ y_{n+1} = \frac{y_n}{2} + \frac{y_n^2}{4} + \frac{x_n y_n}{8}, \end{cases} \tag{5.9}$$

find the fixed points and analyze their stability.

Solution:

Fixed points are roots of the system

$$\begin{cases} x = \frac{x}{2} + \frac{x^2}{4} + \frac{xy}{8} \\ y = \frac{y}{2} + \frac{y^2}{4} + \frac{xy}{8} \end{cases}$$

The first equation is satisfied if either $x = 0$ or

$$1 = 1/2 + x/4 + y/8 \quad \Rightarrow \quad y = 8\left(\frac{1}{2} - \frac{1}{4}x\right) = 4 - 2x.$$

The second equation is satisfied if either $y = 0$ or

$$1 = 1/2 + y/4 + x/8 \quad \Rightarrow \quad x = 8\left(\frac{1}{2} - \frac{1}{4}y\right) = 4 - 2y.$$

By combining these conditions we find the following four fixed points:

- First fixed point: $(X_1, Y_1) = (0, 0)$;
- Second fixed point: from $x = 0$ and $x = 4 - 2y$, we obtain $y = 2$, and hence $(X_2, Y_2) = (0, 2)$;
- Third fixed point: from $y = 0$ and $y = 4 - 2x$, we obtain $x = 2$, and hence $(X_3, Y_3) = (2, 0)$;
- Fourth fixed point: solving $y = 4 - 2x$ and $x = 4 - 2y$ gives

  $$4 - 2(4 - 2y) = 4y - 4 \quad \Rightarrow \quad 3y = 4 \quad \Rightarrow \quad y = \frac{4}{3},$$

  and therefore $x = 4 - 2y = \frac{4}{3}$, so $(X_4, Y_4) = (\frac{4}{3}, \frac{4}{3})$.

Now, we need to identify the types of these fixed points. To do so, we find the Jacobian

$$J = \begin{pmatrix} f_x & f_y \\ g_x & g_y \end{pmatrix}.$$

For

$$f(x,y) = \frac{x}{2} + \frac{x^2}{4} + \frac{xy}{8}, \quad \text{and} \quad g(x,y) = \frac{y}{2} + \frac{y^2}{4} + \frac{xy}{8},$$

the partial derivatives are

$$f_x = \frac{1}{2} + \frac{x}{2} + \frac{y}{8}, \quad f_y = \frac{x}{8}, \quad g_x = \frac{y}{8}, \quad g_y = \frac{1}{2} + \frac{y}{2} + \frac{x}{8}.$$

Hence, the Jacobian is

$$J = \begin{pmatrix} \frac{1}{2} + \frac{1}{2}x + \frac{1}{8}y & \frac{1}{8}y \\ \frac{1}{8}y & \frac{1}{2} + \frac{1}{2}y + \frac{1}{8}x \end{pmatrix}.$$

The next task is to find the Jacobian and its eigenvalues at the fixed points.

- At $(X_1, Y_1) = (0,0)$,

  $$J = \begin{pmatrix} \frac{1}{2} + \frac{1}{2} \times 0 + \frac{1}{8} \times 0 & \frac{1}{8} \times 0 \\ \frac{1}{8} \times 0 & \frac{1}{2} + \frac{1}{2} \times 0 + \frac{1}{8} \times 0 \end{pmatrix} = \begin{pmatrix} \frac{1}{2} & 0 \\ 0 & \frac{1}{2} \end{pmatrix}.$$

  The eigenvalues are roots of the characteristic equation

  $$\begin{vmatrix} \frac{1}{2} - \lambda & 0 \\ 0 & \frac{1}{2} - \lambda \end{vmatrix} = 0 \quad \Rightarrow \quad \left(\frac{1}{2} - \lambda\right)\left(\frac{1}{2} - \lambda\right) = 0 \quad \Rightarrow \quad \lambda_1 = \frac{1}{2}, \lambda_2 = \frac{1}{2}.$$

  We have found that $|\lambda_1| < 1$ and $|\lambda_2| < 1$, and therefore $(X_1, Y_1) = (0,0)$ is a stable node.
- At $(X_2, Y_2) = (0,2)$,

  $$J(0,2) = \begin{pmatrix} \frac{3}{4} & 0 \\ \frac{1}{4} & \frac{3}{2} \end{pmatrix}, \quad \lambda_1 = \frac{3}{4}, \lambda_2 = \frac{3}{2}.$$

  So, $|\lambda_1| < 1$ and $|\lambda_2| > 1$, and therefore $(X_2, Y_2) = (0,2)$ is a saddle.
- At $(X_3, Y_3) = (2,0)$,

  $$J(2,0) = \begin{pmatrix} \frac{3}{2} & \frac{1}{4} \\ 0 & \frac{3}{4} \end{pmatrix}, \quad \lambda_1 = \frac{3}{2}, \lambda_2 = \frac{3}{4}.$$

  So, $|\lambda_1| > 1$ and $|\lambda_2| < 1$, and therefore $(X_3, Y_3) = (2,0)$ is a saddle.
- At $(X_4, Y_4) = (\frac{4}{3}, \frac{4}{3})$,

  $$J\left(\frac{4}{3}, \frac{4}{3}\right) = \begin{pmatrix} \frac{4}{3} & \frac{1}{6} \\ \frac{1}{6} & \frac{4}{3} \end{pmatrix}, \quad \lambda_1 = \frac{3}{2}, \lambda_2 = \frac{7}{6}.$$

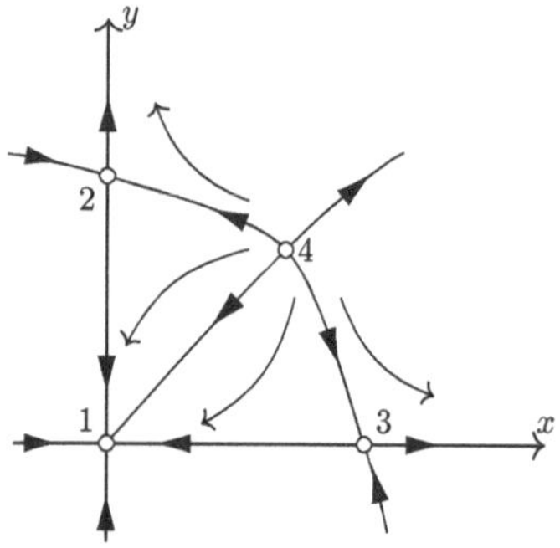

Figure 5.2: Phase portrait for the system (5.9). Fixed points: 1—stable node at (0, 0), 2 and 3 saddles at (0, 2) and (2, 0), 4—unstable node at (4/3, 4/3). Orbits do not flip and (qualitatively) follow shown flows.

Both eigenvalues are greater than 1, and therefore $(X_4, Y_4) = (4/3, 4/3)$ is an unstable node or repeller. An illustration of the gathered material is presented in Figure 5.2. We see that the heteroclinic orbits connecting the unstable node with the saddles divide the phase plane into two regions: trajectories in one region converge to the stable node $(0, 0)$, while those in the other region diverge to infinity.

5.3 2D maps with parameter

Autonomous two-dimensional flow with a single parameter is represented by the following system:

$$\begin{cases} x_{n+1} = f(x_n, y_n, c), \\ y_{n+1} = g(x_n, y_n, c), \end{cases} \tag{5.10}$$

where $f(x, y, c)$ and $g(x, y, c)$ are differentiable functions with respect to the variables x and y and the parameter c. In previous chapters, we have already considered systems with parameters and introduced a few types of bifurcations associated with the way in which the number and stability of fixed points change with variation of the parameter value.

We have identified four types of bifurcations in 1D maps, namely, fold, transcritical, pitchfork, and flip bifurcations. All of them are also observed in 2D maps. Variation of the parameter results in changes to the Jacobian at fixed points. The type of fixed point changes in accordance with the changes in the eigenvalues of the Jacobian. Fold, transcritical, and pitchfork bifurcations are associated with the event when the eigenvalues of the Jacobian at a fixed point are real and one of them becomes equal to one. A flip bifurcation takes place when, under a certain parameter value, one of the eigenvalues is equal to negative one.

There is a fifth type of bifurcation in 2D maps which does not occur in 1D maps. When the eigenvalues of the Jacobian at the fixed point are complex with magnitude equal to one, we observe an analogue of the Hopf bifurcation which, in the case of a

2D map, is called a *Neimark–Sacker bifurcation*. It is associated with the formation of a closed loop in phase space such that there are orbits which stay on it. These orbits represent stable, repeating patterns of either periodic or quasiperiodic motions. The points belonging to all these orbits form a so-called *closed invariant loop*. The term *invariant* refers to the fact that once an orbit hits the curve, it stays on it forever, i. e., the curve maps onto itself. The orbits neighboring this loop are either converging to or diverging from it (i. e., the loop can be stable or unstable).

Example 5.2. Consider the two-dimensional dynamical system defined by the map

$$\begin{cases} x_{n+1} = x_n^2 - y_n^2 + a, \\ y_{n+1} = 3x_n y_n, \end{cases} \tag{5.11}$$

with nonnegative parameter a ($a \geq 0$).

(i) Show that the circle of radius r centered at the origin is mapped, under one iteration of the system, into an ellipse.

(ii) Find all fixed points of the system and classify their types.

(iii) List all bifurcations that occur in this system.

Solution:

(i) Let (x_n, y_n) lie on the circle of radius r centred at the origin

$$x_n = r\cos\theta, \quad y_n = r\sin\theta.$$

Then, one step of the map gives

$$\begin{aligned} x_{n+1} &= x_n^2 - y_n^2 + a = r^2(\cos^2\alpha - \sin^2\alpha) = r^2\cos(2\theta) + a, \\ y_{n+1} &= 3x_n y_n = 3r^2\cos\alpha\sin\alpha = \frac{3}{2}r^2\sin(2\theta). \end{aligned}$$

This is the parametric equation of an ellipse centered at $(a, 0)$ with semi-axes r^2 in the x-direction and $1.5r^2$ in the y-direction. Eliminating θ yields

$$\left(\frac{x_{n+1} - a}{r^2}\right)^2 + \left(\frac{y_{n+1}}{1.5r^2}\right)^2 = \cos^2(2\theta) + \sin^2(2\theta) = 1,$$

i. e., the circle is mapped to an ellipse.

(ii) Fixed points are roots of the algebraic system:

$$\begin{aligned} x &= x^2 - y^2 + a, \\ y &= 3xy. \end{aligned}$$

From the second equation, $y(1 - 3x) = 0$, so either $y = 0$ or $x = \frac{1}{3}$. Consider $y = 0$ such that

$$x = x^2 + a \quad \Rightarrow \quad x_{1,2} = \frac{1 \pm \sqrt{1 - 4a}}{2},$$

which exists only if $a < \frac{1}{4}$. Thus, for $a < \frac{1}{4}$, we obtain two fixed points

$$P_{1,2} = \left(\frac{1 \pm \sqrt{1-4a}}{2}, 0\right).$$

Considering $x = \frac{1}{3}$,

$$\frac{1}{3} = \frac{1}{9} - y^2 + a \quad \Rightarrow \quad y^2 = a - \frac{2}{9} \quad \Rightarrow \quad y_{1,2} = \pm\sqrt{a - \frac{2}{9}}.$$

These solutions exist when $a > \frac{2}{9}$. Thus, two further fixed points are

$$P_{3,4} = \left(\frac{1}{3}, \pm\sqrt{a - \frac{2}{9}}\right).$$

Note that for $\frac{2}{9} < a < \frac{1}{4}$ all four fixed points coexist.
Stability of the fixed points is determined by the eigenvalues of the Jacobian matrix

$$J = \begin{pmatrix} 2x & -2y \\ 3y & 3x \end{pmatrix}.$$

For the first pair of fixed points (with $y = 0$),

$$J(x, 0) = \begin{pmatrix} 2x & 0 \\ 0 & 3x \end{pmatrix}.$$

Eigenvalues for the fixed point $P_1 = (\frac{1+\sqrt{1-4a}}{2}, 0)$ are $\lambda_1 = 1 + \sqrt{1-4a}$ and $\lambda_2 = \frac{3}{2}(1 + \sqrt{1-4a})$. Both eigenvalues satisfy $\lambda_1 > 1$ and $\lambda_2 > 1$ for $a < \frac{1}{4}$. Therefore, the fixed point P_1 is an *unstable node*.
Eigenvalues for the fixed point $P_2 = (\frac{1-\sqrt{1-4a}}{2}, 0)$ are $\lambda_1 = 1 - \sqrt{1-4a}$ and $\lambda_2 = \frac{3}{2}(1 - \sqrt{1-4a})$.
If $0 < a < \frac{2}{9}$, then $0 < \lambda_1 < 1$.
If for λ_2 we require $|\lambda_2| < 1$, then,

$$-1 < \frac{3}{2}(1 - \sqrt{1-4a}) < 1 \quad \Rightarrow \quad -\frac{2}{3} < \sqrt{1-4a} - 1 < \frac{2}{3}.$$

This gives

$$\frac{1}{3} < \sqrt{1-4a} < \frac{5}{3}.$$

Since $\sqrt{1-4a} \le 1$ for $a \ge 0$, the upper bound is automatically satisfied, and the lower bound yields

$$\sqrt{1-4a} > \frac{1}{3} \quad \Rightarrow \quad a < \frac{2}{9}.$$

Thus, the fixed point P_2 is a *stable node* when $a < \frac{2}{9}$ and becomes a *saddle* for $\frac{2}{9} < a < \frac{1}{4}$.

For the second pair of fixed points (with $x = 1/3$),—the Jacobian is

$$J(P_{3,4}) = \begin{pmatrix} 2/3 & \mp 2\sqrt{a - \frac{2}{9}} \\ \pm 3\sqrt{a - \frac{2}{9}} & 1 \end{pmatrix}.$$

At these points,

$$\operatorname{tr} J = \frac{5}{3}, \quad \det J = \frac{2}{3} + 6\left(a - \frac{2}{9}\right) = 6a - \frac{2}{3}.$$

The eigenvalues are roots of $\lambda^2 - \operatorname{tr} J\lambda + \det J = 0$. Their discriminant is

$$\Delta_\lambda = (\operatorname{tr} J)^2 - 4\det J = \frac{25}{9} - 4\left(6a - \frac{2}{3}\right) = \frac{1}{9} - 24\left(a - \frac{2}{9}\right).$$

By examining the sign of the discriminant, we conclude

- For $a \in (\frac{2}{9}, \frac{49}{216})$, the eigenvalues are real and satisfy $|\lambda_{1,2}| < 1$, so the fixed points are *stable nodes.*
- For $a \in (\frac{49}{216}, \frac{5}{18})$, the eigenvalues are complex conjugates with modulus smaller than one, hence the fixed points are *stable spirals.*
- At $a = \frac{5}{18}$, the modulus of the complex conjugate eigenvalues becomes equal to one, after which the fixed points $P_{3,4}$ lose stability and become *unstable spirals.*

(iii) Bifurcations taking place in the system shown in Figure 5.3. They include

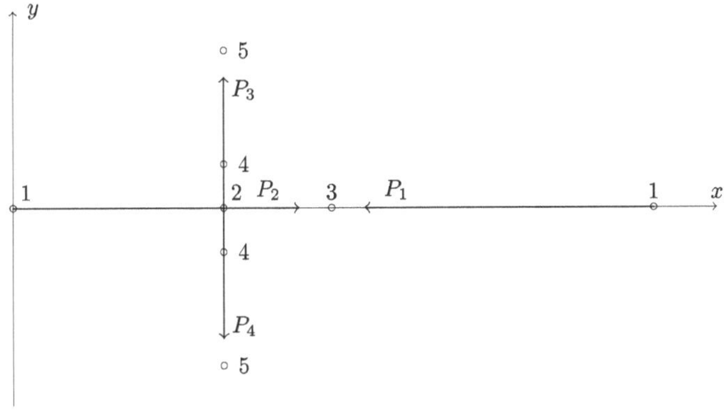

Figure 5.3: Illustration of the dynamical regimes occurring in the system (5.11). When $a = 0$, there are two stable nodes, $P_1 = (0, 1)$ and $P_2 = (0, 0)$ (marked as positions 1). As the parameter a increases, they move towards each other, and at $a = 1/4$ they merge at $P_1 = P_2 = 1/2$ (a saddle-node bifurcation, marked as position 3) and subsequently disappear. At $a = 2/9$, a supercritical pitchfork bifurcation occurs at the location of $P_2 = (1/3, 0)$ (marked as position 2): P_2 becomes a saddle, and two stable nodes, P_3 and P_4, appear. With further increase of a, these two fixed points transform into stable spirals (marked as positions 4), and later into unstable spirals through a Neimark–Sacker bifurcation (marked as positions 5).

- *Fold (saddle-node) bifurcation* at $a = \frac{1}{4}$: As a increases, the two on-axis fixed points P_1 and P_2 coalesce at $(\frac{1}{2}, 0)$ and annihilate.
- *Pitchfork bifurcation* at $a = \frac{2}{9}$: At this parameter value, the fixed point P_2 located at $(\frac{1}{3}, 0)$ loses stability. Simultaneously, two fixed points $P_{3,4}$ emerge at this point, reflecting the pitchfork structure of the bifurcation.
- *Neimark–Sacker bifurcation* at $a = \frac{5}{18}$: Fixed points $P_{3,4}$ transform from stable to unstable spirals (possibly) with formation of an invariant closed loop.

5.4 Hénon map

As we already know, chaotic dynamics can be observed in 1D maps, and the main route to chaos is the so-called *period-doubling cascade*. Chaos is also observed in 2D maps and is commonly demonstrated using the so-called *Hénon map*, which is defined by the following system:

$$\begin{cases} x_{n+1} = a - x_n^2 + by_n, \\ y_{n+1} = x_n, \end{cases}$$

where a and $b \neq 0$ are parameters. The model was introduced by Michel Hénon in 1976 as a simplification of Lorenz model, which we studied in the previous chapter. In this section, we aim to illustrate the formation of chaos in the Hénon map.

Preliminary notes

1. The Hénon map can be reformulated in terms of a single variable.
   Since $y_{n+1} = x_n$, we can replace y_n with x_{n-1} and obtain an equivalent second-order equation in one variable as follows:

$$x_{n+1} = a - x_n^2 + bx_{n-1}.$$

2. The Hénon map is invertible.
   To find the inverse map, we define x_n and y_n in terms of x_{n+1} and y_{n+1}. From the second equation, $x_n = y_{n+1}$, and from the first: $y_n = (x_{n+1} - a + x_n^2)/b = (x_{n+1} - a + ay_{n+1}^2)/b$. Thus,

$$\begin{cases} x_n = y_{n+1}, \\ y_n = \frac{x_{n+1} - a + y_{n+1}^2}{b}. \end{cases}$$

   After inverting time, i. e., replacing n by $n+1$ and $n+1$ by n, we obtain the inverse map

$$\begin{cases} x_{n+1} = y_n, \\ y_{n+1} = \frac{x_n - a + y_n^2}{b}. \end{cases}$$

The Hénon map contains two parameters, which makes its analysis more difficult. For simplicity, we fix one of the parameters, namely, $b = 0.4$, and reduce it to a map with a single parameter:

$$\begin{cases} x_{n+1} = a - x_n^2 + 0.4y_n, \\ y_{n+1} = x_n. \end{cases}$$

We will analyze the dynamics of the Hénon map, paying special attention to the bifurcations that occur when the value of the parameter a is varied.

The fixed points of the Hénon map are the roots of the algebraic system:

$$\begin{cases} x = a - x^2 + 0.4y, \\ y = x, \end{cases}$$

or equivalently,

$$x = a - x^2 + 0.4x.$$

The solutions of this equation are given by

$$x_{1,2} = -0.3 \pm \sqrt{0.09 + a}.$$

And since $y = x$, we also have $y_{1,2} = -0.3 \pm \sqrt{0.09 + a}$. Conclusions from these formulas follow:

- If $a < -0.09$, then there are no fixed points (as $0.09 + a < 0$).
- If $a > -0.09$, then there are two fixed points, $P_1 = (x_1, y_1)$, where $x_1 = y_1 = -0.3 + \sqrt{0.09 + a}$, and $P_2 = (x_2, y_2)$, where $x_1 = y_1 = -0.3 - \sqrt{0.09 + a}$.
- At $a = -0.09$ the two fixed points merge. This is a bifurcation point where a *fold* (or *saddle-node*) bifurcation takes place. This is illustrated in Figure 5.4A.

To assess the stability of the fixed points, we need to calculate the Jacobian

$$J(x, y) = \begin{pmatrix} f'_x & f'_y \\ g'_x & g'_y \end{pmatrix} = \begin{pmatrix} -2x & 0.4 \\ 1 & 0 \end{pmatrix},$$

where the derivatives are taken from the map-generating functions

$$\begin{cases} f(x, y) = a - x^2 + 0.4y, \\ g(x, y) = x. \end{cases}$$

The characteristic equation to find the eigenvalues is

$$\begin{vmatrix} -2x - \lambda & 0.4 \\ 1 & -\lambda \end{vmatrix} = 0 \quad \Rightarrow \quad \lambda^2 + 2x\lambda - 0.4 = 0,$$

and the eigenvalues are

$$\lambda_{1,2} = -x \pm \sqrt{x^2 + 0.4}.$$

When $a = -0.09$ (the bifurcation value), the fixed point is located at

$$x = y = -0.3 \pm \sqrt{0.09 + (-0.09)} = -0.3.$$

The eigenvalues of the Jacobian at this point are

$$\lambda_{1,2} = -x \pm \sqrt{x^2 + 0.4} = -(-0.3) \pm \sqrt{(-0.3)^2 + 0.4} = 0.3 \pm 0.7.$$

So, $\lambda_1 = 1$ and $\lambda_2 = -0.4$. Thus, when $a = -0.09$ the two fixed points coincide, and the Jacobian at this point has $\lambda_1 = 1$ and $|\lambda_2| < 1$. When a decreases, the fixed points disappear, while when a increases they move apart, being located on the diagonal line $y = x$. The Jacobians and corresponding eigenvalues will be different for these two points, so that λ_1 will be greater than 1 for one of them (saddle) and less than 1 for the other fixed point (stable node).

To determine which fixed point is a saddle and which is a node, let us take $a = 0$ and check the location and stability of the fixed points

$$\Rightarrow \quad x = y = -0.3 \pm \sqrt{0.09 + 0} = -0.3 \pm 0.3,$$
$$\Rightarrow \quad x_1 = y_1 = 0, \quad x_2 = y_2 = -0.6.$$

At $P_1 = (0,0)$ we have

$$\lambda_{1,2} = -x \pm \sqrt{x^2 + 0.4} = 0 \pm \sqrt{0.4} \approx \pm 0.632.$$

Since $|\lambda_{1,2}| < 1$, this fixed point is stable node.

At $P_2 = (-0.6, -0.6)$, we have

$$\lambda_{1,2} = -(-0.6) \pm \sqrt{(-0.6)^2 + 0.4} = 0.6 \pm \sqrt{0.76}.$$

Thus,

$$\lambda_1 \approx 1.472, \quad \lambda_2 \approx -0.272.$$

Here, $\lambda_1 > 1$, so this fixed point is a saddle.

The stability of the two fixed points can also be analysed in the vicinity of $a = -0.09$, where a saddle–node bifurcation takes place. In this case, we use a Taylor expansion to evaluate the eigenvalues at the fixed points. When the value of a is slightly above -0.09, the coordinate x_1 of the fixed point $P_1 = (x_1, y_1)$ lies slightly above -0.3, whereas the coordinate x_2 of the fixed point $P_2 = (x_2, y_2)$ lies slightly below -0.3

$$\left.\begin{aligned} x_1 &= -0.3 + \varepsilon, \\ x_2 &= -0.3 - \varepsilon, \end{aligned}\right\} \quad \varepsilon > 0.$$

While λ_2 is still between −1 and 1 we should check λ_1

$$\lambda_1 = -x + \sqrt{x^2 + 0.4}.$$

For $x = -0.3 + \varepsilon$, $\varepsilon > 0$,

$$\begin{aligned} \lambda_1 &= -(-0.3 + \varepsilon) + \sqrt{(-0.3 + \varepsilon)^2 + 0.4} = 0.3 - \varepsilon + \sqrt{\varepsilon^2 - 0.6\varepsilon + 0.49} \\ &\approx 0.3 - \varepsilon + \sqrt{-0.6\varepsilon + 0.49} \quad (\text{the term } \varepsilon^2 \text{ is neglected}) \\ &= 0.3 - \varepsilon + \sqrt{0.49\left(1 - \frac{0.6}{0.49}\varepsilon\right)} = 0.3 - \varepsilon + 0.7\sqrt{1 - \frac{0.6}{0.49}\varepsilon} \\ &= 0.3 - \varepsilon + 0.7\left(1 - \frac{0.3}{0.49}\varepsilon\right) \quad \left(\text{using } \sqrt{1 - x} \approx 1 - \frac{x}{2}\right) \\ &= 0.3 - \varepsilon + 0.7 - \frac{0.3}{0.7}\varepsilon = 1 - 1.43\varepsilon. \end{aligned}$$

Since $\varepsilon > 0$, we have $\lambda_1 < 1$, and thus $P_1 = (-0.3 + \varepsilon, -0.3 + \varepsilon)$ is a stable node.

The stability of the fixed point P_1 also depends on the eigenvalue λ_2 which, in turn, depends on the location of P_1. The value of λ_2 decreases as the coordinates of P_1 increase, and we can determine the position of P_1 at which λ_2 crosses the value −1, where the fixed point P_1 loses its stability. That is, we seek the value of x such that

$$\lambda_2 = -x - \sqrt{x^2 + 0.4} = -1.$$

This corresponds to the following position of the fixed point $P_1 = (x, x)$, where it becomes unstable:

$$\begin{aligned} \lambda_2 = -1 \quad &\Rightarrow \quad -x - \sqrt{x^2 + 0.4} = -1 \quad \Rightarrow \quad 1 - x = \sqrt{x^2 + 0.4}, \\ &\Rightarrow \quad (1 - x)^2 = x^2 + 0.4 \quad \Rightarrow \quad 1 - 2x + x^2 = x^2 + 0.4, \\ &\Rightarrow \quad 2x = 0.6 \quad \Rightarrow \quad x = 0.3. \end{aligned}$$

Thus, the fixed point P_1 moves along the line $y = x$ upwards as the parameter a increases, and when it reaches the location (0.3, 0.3), it becomes unstable. At this point, λ_1 is still between −1 and 1, and therefore P_1 transforms into a saddle point. As P_1 loses stability when $\lambda_2 = -1$, a flip (period-doubling) bifurcation occurs, and we should expect the emergence of a 2-periodic solution.

To find the corresponding value of a, we impose

$$x = -0.3 + \sqrt{0.09 + a} = 0.3 \quad \Rightarrow \quad \sqrt{0.09 + a} = 0.6 \quad \Rightarrow \quad a = 0.27.$$

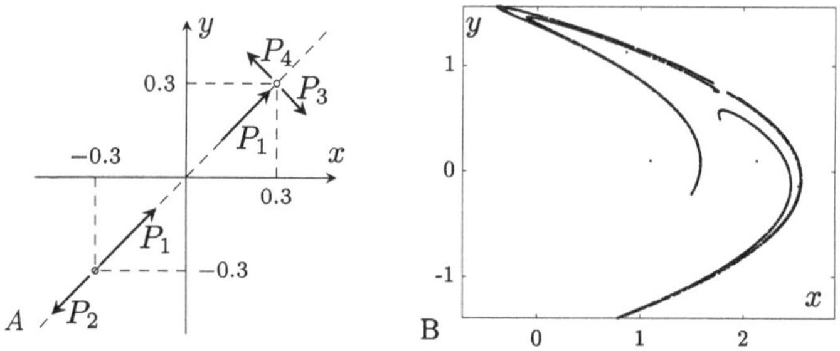

Figure 5.4: Panel A: Illustration of bifurcations in the Hénon map. At $a = -0.09$, there is a saddle-node bifurcation at the point $(-0.3, -0.3)$. As a increases, P_1, which is a stable node, moves upward along the line $x = y$, while P_2, a saddle, moves downward. At $a = 0.27$, the point P_1 reaches the location $(0.3, 0.3)$, where a flip bifurcation occurs. In the second-iterate map, a supercritical pitchfork bifurcation takes place: P_1 becomes a saddle, while two new fixed points, P_3 and P_4, appear as stable nodes. **Panel B:** The strange attractor that forms with a further increase of the parameter a, here shown for $a = 1.1$.

Hence, the fixed point P_1 is *stable* when

$$-0.09 < a < 0.27,$$

becoming unstable at $a = 0.27$ due to the flip bifurcation (illustrated in Figure 5.4A).

To understand what happens at $a = 0.27$, we have to consider the 2-composite map. The 2-periodic solutions correspond to the fixed points of the second iterate

$$\begin{cases} x_{n+2} = f^{(2)}(x_n, y_n) = a - (a - x_n^2 + 0.4y_n)^2 + 0.4x_n, \\ y_{n+2} = g^{(2)}(x_n, y_n) = a - x_n^2 + 0.4y_n. \end{cases}$$

Equivalently, the 2-periodic points are the roots of the algebraic system given by the fourth-order polynomial

$$\begin{cases} x = a - (a - x^2 + 0.4y)^2 + 0.4x, \\ y = a - x^2 + 0.4y. \end{cases}$$

By solving this system, one can show that the stable 2-periodic solution emerges precisely at $a = 0.27$ (illustrated in Figure 5.4A). As the parameter a increases further, the system undergoes successive period-doubling bifurcations, giving rise to a cascade analogous to that observed in the logistic map.

This cascade leads to the creation of stable orbits of period $2, 4, 8, \ldots$, with the 2^n-periodic orbit losing its stability at the value of the parameter a at which a stable 2^{n+1}-periodic orbit arises. Each successive bifurcation occurs as the parameter increases, but within progressively smaller intervals of a. Ultimately, this accumulation of bifurcations results in the onset of chaotic dynamics, associated with the formation of strange attractor, shown in Figure 5.4B.

5.5 Changes in local element areas

The state of the dynamical system (x_n, y_n) described by the two-dimensional map (5.1) can be regarded as a point with position vector $\mathbf{r}_n$ in the x/y-plane, which undergoes a discrete jump at each iteration of the map. During the n-th iteration, the point $\mathbf{r}_n$ is mapped into the point $\mathbf{r}_{n+1}$, and points in the neighborhood of $\mathbf{r}_n$ are mapped into points in the neighborhood of $\mathbf{r}_{n+1}$.

Furthermore, all points comprising the area A_n in the neighborhood of $\mathbf{r}_n$ are mapped into points forming the area A_{n+1} in the neighborhood of point $\mathbf{r}_{n+1}$ (see Figure 5.5A), and we can ask (similarly to what we did with flows in the previous chapter) whether A_{n+1} is larger, smaller, or equal to A_n.

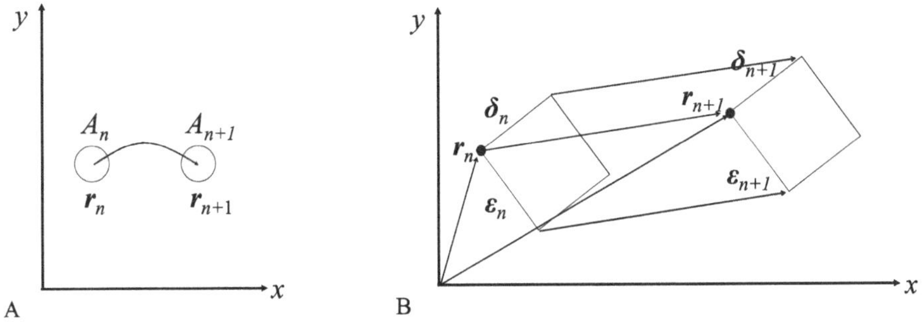

Figure 5.5: A: Local change of an area element under one iteration of the map. **B:** Illustration to the analysis of the local change of an area element shaped as a parallelogram.

Theorem 5.3. *The change in the size of the local area element A_n at the point $\mathbf{r}_n$ under one iteration of the map is governed by the Jacobian determinant*

$$A_{n+1} = \left|\det J(\mathbf{r}_n)\right| A_n,$$

where $J(\mathbf{r}_n)$ denotes the Jacobian matrix evaluated at the point $\mathbf{r}_n$.

Proof. Let us prove the theorem for a neighborhood shaped as a parallelogram. We know that the area of a parallelogram built on the vectors $\boldsymbol{\varepsilon}$ and $\boldsymbol{\delta}$ is given by

$$A = |\boldsymbol{\varepsilon} \times \boldsymbol{\delta}|,$$

where $|\boldsymbol{\varepsilon} \times \boldsymbol{\delta}|$ denotes the magnitude of the cross product

$$\boldsymbol{\varepsilon} \times \boldsymbol{\delta} = \begin{vmatrix} \mathbf{i} & \mathbf{j} & \mathbf{k} \\ \varepsilon_x & \varepsilon_y & 0 \\ \delta_x & \delta_y & 0 \end{vmatrix} = (\varepsilon_x \delta_y - \varepsilon_y \delta_x)\mathbf{k},$$

or explicitly,

$$A = |\boldsymbol{\varepsilon} \times \boldsymbol{\delta}| = |\varepsilon_x \delta_y - \varepsilon_y \delta_x|.$$

Now, let A_n be a parallelogram with one edge at $\boldsymbol{r}_n$ and sides given by small vectors $\boldsymbol{\epsilon}_n$ and $\boldsymbol{\delta}_n$ (see Figure 5.5B). Its image (in the linear approximation) is a parallelogram A_{n+1} built on vectors $\boldsymbol{\epsilon}_{n+1}$ and $\boldsymbol{\delta}_{n+1}$, which we should determine. For the areas of these two parallelograms, we have

$$\begin{aligned} A_n &= |\varepsilon_{n,x}\delta_{n,y} - \varepsilon_{n,y}\delta_{n,x}|, \\ A_{n+1} &= |\varepsilon_{n+1,x}\delta_{n+1,y} - \varepsilon_{n+1,y}\delta_{n+1,x}|. \end{aligned}$$

The image of the point $\boldsymbol{r}_n$ is $\boldsymbol{r}_{n+1}$ as follows:

$$\boldsymbol{r}_{n+1} = \begin{pmatrix} x_{n+1} \\ y_{n+1} \end{pmatrix} = \begin{pmatrix} f(x_n, y_n) \\ g(x_n, y_n) \end{pmatrix} = \begin{pmatrix} f(\boldsymbol{r}_n) \\ g(\boldsymbol{r}_n) \end{pmatrix}.$$

To find the vector $\boldsymbol{\varepsilon}_{n+1}$, we note that the point $\boldsymbol{r}_{n+1} + \boldsymbol{\varepsilon}_{n+1}$ is the image of $\boldsymbol{r}_n + \boldsymbol{\varepsilon}_n$ and is given by

$$\begin{aligned} \boldsymbol{r}_{n+1} + \boldsymbol{\varepsilon}_{n+1} &= \begin{pmatrix} f(\boldsymbol{r}_n + \boldsymbol{\varepsilon}_n) \\ g(\boldsymbol{r}_n + \boldsymbol{\varepsilon}_n) \end{pmatrix} = \begin{pmatrix} f(x_n + \varepsilon_{n,x}, y_n + \varepsilon_{n,y}) \\ g(x_n + \varepsilon_{n,x}, y_n + \varepsilon_{n,y}) \end{pmatrix} \\ &= \begin{pmatrix} f(x_n, y_n) + f'_x \cdot \varepsilon_{n,x} + f'_y \cdot \varepsilon_{n,y} \\ g(x_n, y_n) + g'_x \cdot \varepsilon_{n,x} + g'_y \cdot \varepsilon_{n,y} \end{pmatrix} \\ &= \boldsymbol{r}_{n+1} + J(\boldsymbol{r}_n) \cdot \boldsymbol{\varepsilon}_n. \end{aligned}$$

A similar result holds for $\boldsymbol{\delta}_{n+1}$. Thus,

$$\begin{cases} \boldsymbol{\varepsilon}_{n+1} = J(\mathbf{r}_n)\boldsymbol{\varepsilon}_n, \\ \boldsymbol{\delta}_{n+1} = J(\mathbf{r}_n)\boldsymbol{\delta}_n. \end{cases}$$

Coming back to A_{n+1}, we have

$$\begin{aligned} A_{n+1} &= |\varepsilon_{n+1,x}\delta_{n+1,y} - \varepsilon_{n+1,y}\delta_{n+1,x}| \\ &= (f'_x \cdot \varepsilon_{n,x} + f'_y \cdot \varepsilon_{n,y}) \cdot (g'_x \cdot \delta_{n,x} + g'_y \cdot \delta_{n,y}) - (g'_x \cdot \varepsilon_{n,x} + g'_y \cdot \varepsilon_{n,y}) \cdot (f'_x \cdot \delta_{n,x} + f'_y \cdot \delta_{n,y}) \\ &= \det J(\mathbf{r}_n) \cdot \underbrace{(\varepsilon_{n,x} \cdot \delta_{n,y} - \varepsilon_{n,y} \cdot \delta_{n,x})}_{A_n} \\ &= \det J(\mathbf{r}_n) \cdot A_n, \end{aligned}$$

which completes the proof for parallelogram neighborhoods. □

The prove can be extended to the area of any (smooth) shape as such area can be split into collection of parallelograms. Thus, the Jacobian $J(x, y)$ measures the local expansion or contraction of area elements with the following summary:

- If $|\det J(x, y)| < 1$, the map is locally *contracting*, and areas shrink under iteration.
- If $|\det J(x, y)| > 1$, the map is locally *expanding*, and areas grow under iteration.
- If $|\det J(x, y)| = 1$, the map is locally *area-preserving*.

The obtained result on local changes of area elements can be extended to changes over entire orbits since the area A_n can be expressed in terms of A_0 as

$$\begin{aligned} A_n &= |\det J(\mathbf{r}_{n-1})|A_{n-1} = |\det J(\mathbf{r}_{n-1})||\det J(\mathbf{r}_{n-2})|A_{n-2} \\ &= |\det J(\mathbf{r}_{n-1})||\det J(\mathbf{r}_{n-2})| \dots |\det J(\mathbf{r}_0)|A_0. \end{aligned}$$

Alternatively,

$$A_n = |\det(\mathbb{J}_n)|A_0, \tag{5.12}$$

where

$$\mathbb{J}_n = J(\mathbf{r}_{n-1})J(\mathbf{r}_{n-2}) \dots J(\mathbf{r}_0).$$

If we start with an initial point $\mathbf{r}_0$ and its neighborhood of area A_0, then the sequence of areas $\{A_n\}$ may increase, decrease, or remain unchanged over the course of iterations for a given orbit.

Definition 5.10. If, for a given map, the changes in element areas are qualitatively the same for all orbits, then we can introduce definitions of maps in analogy with those of multidimensional flows:

- If $\{A_n\}$ is an increasing sequence and $A_n \to \infty$ as $n \to \infty$, the map is called **area increasing**.
- If A_n remains unchanged, or more generally if A_n/A_0 tends to a finite constant as $n \to \infty$, the map is called **area conserving** or a **conservative map**.
- If $\{A_n\}$ is a decreasing sequence and $A_n \to 0$ as $n \to \infty$, the map is called **area decreasing** or a **dissipative map**.

As we know, chaotic orbits are sensitive to initial conditions and are bounded. Sensitivity to initial conditions may occur in all three types of maps, as will be shown in the next section. Maps of all three types allow both bounded and unbounded orbits. In dissipative and conservative maps, the formation of unbounded orbits is associated with the change (stretching) of the local area shape. In Figure 5.6 this is illustrated for a conservative map. On the other hand, even in area-increasing maps the orbits can be bounded due to the folding effect, which we have seen in the logistic map. We conclude that in all three types of maps a chaotic regime can be observed.

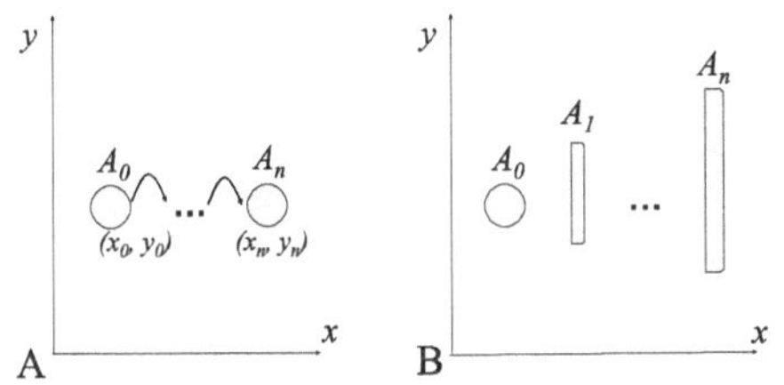

Figure 5.6: Two examples of conservative systems. **A:** Shape and area are preserved; **B:** Area is preserved by the shape is changing.

5.6 Lyapunov exponent

As we have seen from the consideration of one-dimensional maps, sensitivity to initial conditions is associated with the divergence of nearby orbits and is characterised by the Lyapunov exponent. Let us now look at this characteristic of orbits in two-dimensional maps. Consider two nearby orbits

$$\boldsymbol{r}_0 \to \boldsymbol{r}_1 \to \boldsymbol{r}_2 \to \cdots \to \boldsymbol{r}_n \to \cdots \quad \text{and}$$
$$\boldsymbol{r}_0 + \Delta\boldsymbol{r}_0 \to \boldsymbol{r}_1 + \Delta\boldsymbol{r}_1 \to \boldsymbol{r}_2 + \Delta\boldsymbol{r}_2 \to \cdots \to \boldsymbol{r}_n + \Delta\boldsymbol{r}_n \to \cdots,$$

where $\boldsymbol{r}_n = (x_n, y_n)$ represents one of the orbits, and $\Delta\boldsymbol{r}_n = (\varepsilon_n, \delta_n)$ gives the relative position of the second orbit with respect to the first one (see Figure 5.7).

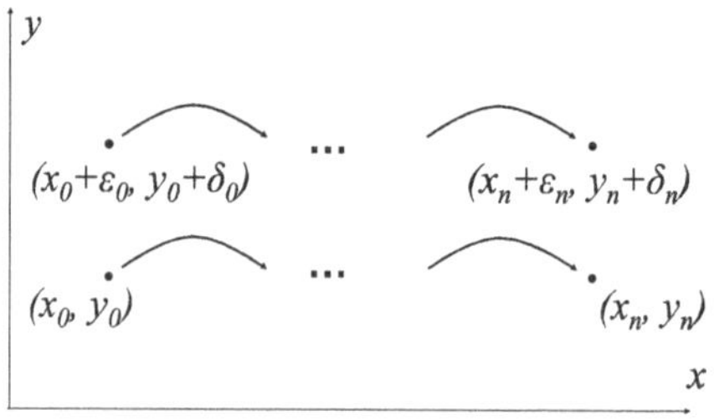

Figure 5.7: Illustration of two nearby orbits.

Let us assume that initially the two orbits are very close to each other, and the pair $(\varepsilon_0, \delta_0)$ can be represented by two infinitesimals. If $\Delta\boldsymbol{r}_n$ increases, the two orbits will diverge, and if $\Delta\boldsymbol{r}_n$ decreases, they will converge. The distance between the orbits changes from $\sqrt{\varepsilon_0^2 + \delta_0^2}$ to $\sqrt{\varepsilon_n^2 + \delta_n^2}$, and we can state that

- If $\sqrt{\varepsilon_n^2 + \delta_n^2} \to 0$ as $n \to \infty$, the system is nonsensitive to initial conditions.
- If $\sqrt{\varepsilon_n^2 + \delta_n^2} \to \infty$ as $n \to \infty$, the system is sensitive to initial conditions.

Using the result obtained in the previous section, we can write

$$\Delta\boldsymbol{r}_n = J(\boldsymbol{r}_{n-1})\Delta\boldsymbol{r}_{n-1}.$$

A similar formula links $\Delta\boldsymbol{r}_{n-1}$ with $\Delta\boldsymbol{r}_{n-2}$, and so on, and thus this is easily extended to

$$\Delta\boldsymbol{r}_n = J(\mathbf{r}_{n-1})J(\mathbf{r}_{n-2})\ldots J(\mathbf{r}_0)\Delta\boldsymbol{r}_0 = \mathbb{J}_n\Delta\boldsymbol{r}_0,$$

or,

$$\begin{pmatrix}\varepsilon_n\\ \delta_n\end{pmatrix} = \mathbb{J}_n\begin{pmatrix}\varepsilon_0\\ \delta_0\end{pmatrix},$$

where

$$\mathbb{J}_n = J(\mathbf{r}_{n-1})J(\mathbf{r}_{n-2})\ldots J(\mathbf{r}_0). \tag{5.13}$$

Now, let $(\Lambda_1, \mathbf{b}_1)$ and $(\Lambda_2, \mathbf{b}_2)$ be the eigensolutions of the matrix $\mathbb{J}_n$. Then, if the two eigenvectors are linearly independent, we can find scalars c_1 and c_2 such that

$$\begin{pmatrix} \varepsilon_0 \\ \delta_0 \end{pmatrix} = c_1\mathbf{b}_1 + c_2\mathbf{b}_2.$$

Furthermore, for the difference between the orbits after n iterations

$$\begin{pmatrix} \varepsilon_n \\ \delta_n \end{pmatrix} = c_1\Lambda_1\mathbf{b}_1 + c_2\Lambda_2\mathbf{b}_2.$$

The evolution of $\left(\begin{smallmatrix} \varepsilon_n \\ \delta_n \end{smallmatrix}\right)$ is determined by Λ_1 and Λ_2, which also vary as n increases. Assume that

$$\begin{cases} |\Lambda_1| \to e^{nLE_1}, & \text{as } n \to \infty, \\ |\Lambda_2| \to e^{nLE_2}, & \text{as } n \to \infty \end{cases} \tag{5.14}$$

where LE_1 and LE_2 are the so-called *Lyapunov exponents*. The assumption (5.14) can be justified by the following reasoning. If, for example, $|\Lambda_1|$ grows more slowly than exponentially, then $LE_1 = 0$. If it grows faster than exponentially, then $LE_1 = \infty$. And if we have exponential growth, then the Lyapunov exponent LE_1 is a finite number.

From (5.14) we can obtain the following explicit formulae for the Lyapunov exponents:

$$\begin{cases} LE_1 = \lim_{n\to\infty}(\frac{1}{n}\ln|\Lambda_1|), \\ LE_2 = \lim_{n\to\infty}(\frac{1}{n}\ln|\Lambda_2|). \end{cases} \tag{5.15}$$

Lyapunov exponents give a clear indication of the behavior of nearby orbits since

1. If $LE_1 < 0$ and $LE_2 < 0$, then $\Lambda_1 \to 0$ and $\Lambda_2 \to 0$, and correspondingly $\varepsilon_n \to 0$ and $\delta_n \to 0$. Thus, the system is insensitive to initial conditions.
2. If $LE_1 > 0$ and $LE_2 > 0$, then $\Lambda_1 \to \infty$ and $\Lambda_2 \to \infty$, and correspondingly $\varepsilon_n \to \infty$ and $\delta_n \to \infty$. In this case, we have sensitivity to initial conditions, however the solutions are unbounded because the map is area-expanding. Indeed,

   $$\det(\mathbb{J}) = |\Lambda_1 \cdot \Lambda_2| = e^{nLE_1} \cdot e^{nLE_2} = e^{n(LE_1+LE_2)} \to \infty \quad \text{as } n \to \infty,$$

   and thus, according to (5.12), $A_n \to \infty$. As the solutions are unbounded, we can not have chaos in the system.
3. If $LE_1 > 0$ and $LE_2 < 0$, then we have sensitivity to initial conditions, and if the solutions are bounded, we will have chaos. The solutions can only be bounded if the system is not expanding, i. e.,

   $$\det(\mathbb{J}) = |\Lambda_1 \cdot \Lambda_2| = e^{nLE_1} \cdot e^{nLE_2} = e^{n(LE_1+LE_2)} \nrightarrow \infty \quad \text{as } n \to \infty,$$

   and thus, according to (5.12), $A_n \nrightarrow \infty$ only if $LE_1 + LE_2 \le 0$.

The known chaotic regimes in natural systems are mostly associated with dissipative systems, where $LE_1 + LE_2 < 0$, although conservative systems with $LE_1 + LE_2 = 0$ can also exhibit chaos.

As Λ_1 and Λ_2 are eigenvalues of the matrix $\mathbb{J}_n$, defined as the product of Jacobians along the orbit (5.13), we can arrange the corresponding pairs of Jacobian eigenvalues so that

$$\Lambda_1 \approx \lambda_{0,1}\lambda_{1,1}\lambda_{2,1}\cdots\lambda_{n-1,1},$$

and

$$\Lambda_2 \approx \lambda_{0,2}\lambda_{1,2}\lambda_{2,2}\cdots\lambda_{n-1,2},$$

where $\lambda_{i,1}$ and $\lambda_{i,2}$ are the eigenvalues of $J(\mathbf{r}_i)$. The approximation becomes an exact equality when all Jacobians along the orbit commute since, in that case, the eigenvalues of the product are exactly the products of the eigenvalues.

Let

$$\bar{\lambda}_1 = |\lambda_{0,1}\lambda_{1,1}\lambda_{2,1}\cdots\lambda_{n-1,1}|^{\frac{1}{n}},$$

which is the geometric average of the eigenvalues of the Jacobian matrices along the orbit. Then,

$$|\Lambda_1| \approx \bar{\lambda}_1^n = e^{n\ln(\bar{\lambda}_1)},$$

and thus,

$$LE_1 \approx \ln(\bar{\lambda}_1).$$

Similarly,

$$LE_2 \approx \ln(\bar{\lambda}_2),$$

where

$$\bar{\lambda}_2 = |\lambda_{0,2}\lambda_{1,2}\lambda_{2,2}\cdots\lambda_{n-1,2}|^{\frac{1}{n}}.$$

A more commonly used formula for the Lyapunov exponents is

$$LE_1 = \lim_{n\to\infty}\left(\frac{1}{n}\ln(\Lambda_1)\right) = \lim_{n\to\infty}\left(\frac{1}{n}\ln|\lambda_{0,1}\lambda_{1,1}\lambda_{2,1}\cdots\lambda_{n-1,1}|\right) = \lim_{n\to\infty}\left(\frac{1}{n}\sum_{i=0}^{n-1}\ln|\lambda_{i,1}|\right),$$

and similarly,

$$LE_2 = \lim_{n\to\infty}\left(\frac{1}{n}\sum_{i=0}^{n-1}\ln|\lambda_{i,2}|\right).$$

Example 5.3. Find the Lyapunov exponents for the linear map.

Solution:
For a linear system, the Jacobian is the same throughout the phase plane, so $J_0 = J_1 = \cdots = J_{n-1}$. Hence,

$$J = J_{n-1}J_{n-2}\cdots J_0 = J_0^n \quad \Rightarrow \quad \begin{cases} \Lambda_1 = \lambda_1^n, \\ \Lambda_2 = \lambda_2^n, \end{cases}$$

where λ_1 and λ_2 are the eigenvalues of J_0.

Correspondingly,

$$LE_1 = \lim_{n\to\infty}\left(\frac{1}{n}\ln(\Lambda_1)\right) = \lim_{n\to\infty}\left(\frac{1}{n}\ln|\lambda_1|^n\right) = \ln|\lambda_1|.$$

Similarly,

$$LE_2 = \lim_{n\to\infty}\left(\frac{1}{n}\ln(\Lambda_2)\right) = \lim_{n\to\infty}\left(\frac{1}{n}\ln|\lambda_2|^n\right) = \ln|\lambda_2|.$$

If the fixed point of the linear system is stable, then $|\lambda_1| < 1$ and $|\lambda_2| < 1$, and therefore $LE_1 < 0$ and $LE_2 < 0$.

Example 5.4. Find the sum of the Lyapunov exponents for the dynamical system under the assumption that the determinant of the Jacobian is the same throughout the phase space.

Solution:
Since $\det(J)$ is constant, we have

$$\Lambda_1\Lambda_2 = \det(\mathbb{J}_n) = |\det J(\mathbf{r_{n-1}})||\det J(\mathbf{r_{n-2}})|\ldots|\det J(\mathbf{r_0})| = [\det(J)]^n.$$

This gives for the sum of the following Lyapunov exponents:

$$\begin{aligned} LE_1 + LE_2 &= \lim_{n\to\infty}\left(\frac{1}{n}\ln|\Lambda_1|\right) + \lim_{n\to\infty}\left(\frac{1}{n}\ln|\Lambda_2|\right) \\ &= \lim_{n\to\infty}\left(\frac{1}{n}(\ln|\Lambda_1| + \ln|\Lambda_2|)\right) = \lim_{n\to\infty}\left(\frac{1}{n}\ln|\Lambda_1\Lambda_2|\right) \\ &= \lim_{n\to\infty}\left(\frac{1}{n}\ln|\det(J)|^n\right) = \ln|\det(J)|. \end{aligned}$$

This formula confirms that the type of dynamical system, characterized by $\det(J)$, can also be described in terms of the sum of the Lyapunov exponents.

5.7 Summary

In this chapter we have considered autonomous 2D maps given by (5.1) for which solution is represented by a set of points (x_i, y_i), $i = 0, 1, \ldots, n, \ldots$ on the phase plane. Analysis of the dynamics in these maps is associated with finding fixed points (X, Y) given as roots of the algebraic system (5.2), and identifying of their types. Nullclines, which were very useful for the analysis of 2D flows, are not applicable to maps since in the latter the orbits are discrete and not directly associated with the derivatives of the map-generation functions.

The stability and type of a fixed point are defined by the eigenvalues λ_1 and λ_2 of the Jacobian (5.8) at this point as follows:
- If $|\lambda_1| < 1$ and $|\lambda_2| < 1$, the fixed point (X, Y) is asymptotically stable.
- If $|\lambda_1| > 1$ and/or $|\lambda_2| > 1$, the fixed point (X, Y) is unstable.
- If $|\lambda_1| < 1$ and $\lambda_2 = 1$, then in the case of a linear system we would have neutral stability. However, in the case of a nonlinear system, we need to go beyond linearization to determine whether the fixed point is stable or not.

Thus, the fixed points in maps are stable if the eigenvalues of the Jacobian at this point lie inside the unit circle in the complex plane (see Figure 5.8A), while for flows the eigenvalues should lie in the half-space to the left of the vertical axis (see Figure 5.8B).

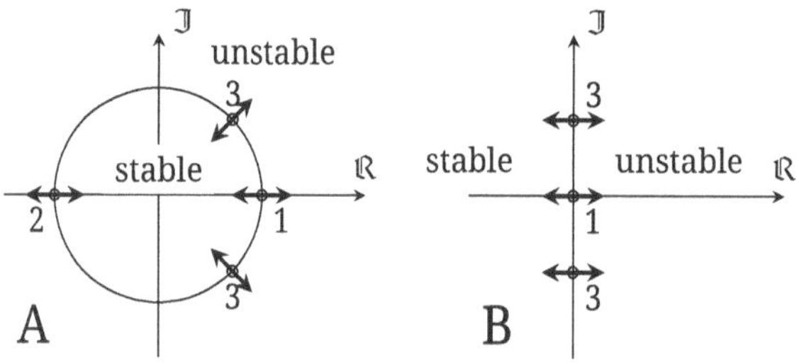

Figure 5.8: Panel A: A fixed point in a 2D map is stable if both eigenvalues of the Jacobian at this point lie inside the unit circle in the complex plane. Saddle-node, transcritical, or pitchfork bifurcations take place when one of the eigenvalues is equal to one (location marked as '1'). A flip bifurcation takes place when one of the eigenvalues is −1 (marked as '2'). Finally, the Neimark–Sacker bifurcation takes place when the eigenvalues are complex conjugates and lie on the unit circle (marked as '3'). **Panel B:** A fixed point in a 2D flow is stable if both eigenvalues of the Jacobian lie to the left of the vertical axis in the complex plane. Saddle-node, transcritical, or pitchfork bifurcations take place when one of the eigenvalues is equal to zero (location marked as '1'). The Hopf bifurcation takes place when the eigenvalues are complex conjugates and lie on the vertical axis (marked as '3').

Sketching the phase portrait of the map is similar to that for the flow and entails
- Plotting all the fixed points;
- Sketching the stable and unstable manifolds for saddle points; and
- Illustrating the direction of the sketched sample orbits.

In maps with a parameter we observe five types of bifurcations. When the eigenvalues of the Jacobian at the fixed point are real and one of them crosses the value 1 (in the course of a smooth change of the parameter), we can have a saddle-node, transcritical, or pitchfork bifurcation (marked as '1' in Figure 5.8A). As we found in the previous chapter, these three bifurcations are observed in flows when one of the eigenvalues crosses the value 0 (marked as '1' in Figure 5.8B). When one of the eigenvalues crosses the value −1, a flip bifurcation is observed (marked as '2' in Figure 5.8A), which does not occur in flows. Finally, if the eigenvalues are complex conjugates and their modulus crosses the value 1 (marked as '3' in Figure 5.8A), a Neimark–Sacker bifurcation is observed in maps. This is the analogue of the Hopf bifurcation observed in flows when the real part of a complex conjugate pair of eigenvalues crosses the value 0 (marked as '3' in Figure 5.8B).

Figure 5.8 illustrates the differences in the results concerning the stability of fixed points and the conditions for bifurcations in 2D flows and 2D maps. Another important difference concerns the formation of chaotic regimes. According to the Poincaré–Bendixson theorem, chaos cannot form in 2D flows, while, as we have seen, chaos arises quite naturally in 2D maps. Importantly, chaos can form in invertible and dissipative 2D maps (we have shown that the Hénon map is invertible and dissipative). The only necessary condition is that at least one of the Lyapunov exponents along the chaotic orbit is positive. In dissipative systems, chaos is associated with the formation of strange attractors, which have a so-called *fractal* structure. Geometrical objects called fractals will be considered in the next chapter.

5.8 Exercises

Problem 5.1. Consider a two-dimensional dynamical system defined by the map

$$\begin{cases} x_{n+1} = 0.5x_n^2 + y_n^2, \\ y_{n+1} = x_n y_n. \end{cases} \tag{5.16}$$

(i) Find all its fixed points and analyse their stability.
(ii) Plot the fixed points on the (x, y)-plane, sketch the associated stable and unstable manifolds, and illustrate the directions of orbits in the vicinity of the fixed points.

Problem 5.2. Consider the dynamical system

$$x_{n+1} = \begin{cases} bx_n, & 0 \leq x_n \leq \frac{1}{2}, \\ \frac{1}{2} + bx_n, & \frac{1}{2} < x_n \leq 1, \end{cases}$$
$$y_{n+1} = 2y_n \pmod 1,$$

where $0 < b \leq 1/2$.
(i) Show that the system is area preserving for $b = 1/2$.

(ii) Sketch the image of the line $y = x$ after one iteration.
(iii) Obtain the Jacobian matrix and show that the Lyapunov exponents are $\ln 2$ in the y-direction and $\ln b$ in the x-direction.

Problem 5.3. Consider a two-dimensional dynamical system defined by the map

$$\begin{cases} x_{n+1} = 1 + y_n - ax_n^2, \\ y_{n+1} = x_n, \end{cases}$$

where $a \in \mathbb{R}$ is a parameter.
(i) Show that this map is invertible by finding the inverse map.
(ii) Find all fixed points of this system and analyse their stability. What kind of bifurcation takes place in this system when $a = 0$?

Problem 5.4. Consider the dynamical system given by

$$\begin{cases} x_{n+1} = y_n + 1 - ax_n^2, \\ y_{n+1} = (1 - B)x_n, \end{cases}$$

where a and B are real constants and $B > 0$.
(i) Find the fixed points of this system and show that the system has a stable fixed point only when

$$-\frac{1}{4}B^2 < a < \frac{3}{4}B^2.$$

Classify the fixed points in the case $4a = 3B^2$ with $B = 0.5$.
(ii) Assuming that the additional fixed points $(X_{3,4}, Y_{3,4})$ associated with the 2-periodic solution of the given map are such that they are the roots of

$$a^2x^2 - aBx + B^2 - a = 0,$$

find the product of the Jacobians at these fixed points. Then, show that the eigenvalues of this product satisfy

$$\lambda^2 - \left[4a^2X_3X_4 + 2(1 - B)\right]\lambda + (1 - B)^2 = 0.$$

Deduce that when $B = 1$, the system possesses a stable 2-periodic solution precisely when

$$\frac{3}{4} < a < \frac{5}{4}.$$

5.9 Solutions to exercises

Solution to Problem 5.1.

(i) Fixed points are the roots of the algebraic system

$$\begin{cases} x = 0.5x^2 + y^2, \\ y = xy. \end{cases}$$

This system has four roots: $(X_1, Y_1) = (0, 0)$, $(X_2, Y_2) = (2, 0)$, $(X_{3,4}, Y_{3,4}) = (1, \pm\sqrt{1/2})$. Stability of the fixed points is defined by the eigenvalues of the Jacobian matrix

$$J = \begin{pmatrix} x & 2y \\ y & x \end{pmatrix}.$$

At $(X_1, Y_1) = (0, 0)$,

$$J(0, 0) = \begin{pmatrix} 0 & 0 \\ 0 & 0 \end{pmatrix}$$

with eigenvalues $\lambda_1 = \lambda_2 = 0$, both of which have absolute value less than 1, and therefore (X_1, Y_1) is a stable node.
At $(X_2, Y_2) = (2, 0)$,

$$J(2, 0) = \begin{pmatrix} 2 & 0 \\ 0 & 2 \end{pmatrix}$$

with eigenvalues $\lambda_1 = \lambda_2 = 2$, both of which have absolute value greater than 1, and therefore (X_2, Y_2) is an unstable node.
At the two other equilibrium points $(X_{3,4}, Y_{3,4}) = (1, \pm\sqrt{1/2})$,

$$J(1, \pm\sqrt{1/2}) = \begin{pmatrix} 1 & \pm\sqrt{2} \\ \pm\sqrt{1/2} & 1 \end{pmatrix},$$

with eigenvalues $\lambda_1 = 0$ and $\lambda_2 = 2$. As $|\lambda_1| < 1$ and $|\lambda_2| > 1$, these two equilibrium points are saddles.

(ii) The phase plane with the marked fixed points, sketches of the stable and unstable manifolds of the saddles, and illustrative sample flows in different regions separated by heteroclinic orbits is presented in Figure 5.9. As the eigenvalues of the Jacobian at and near the fixed points are positive (or at least close to zero), we can assume that the orbits do not flip and therefore follow the indicated sample flows.

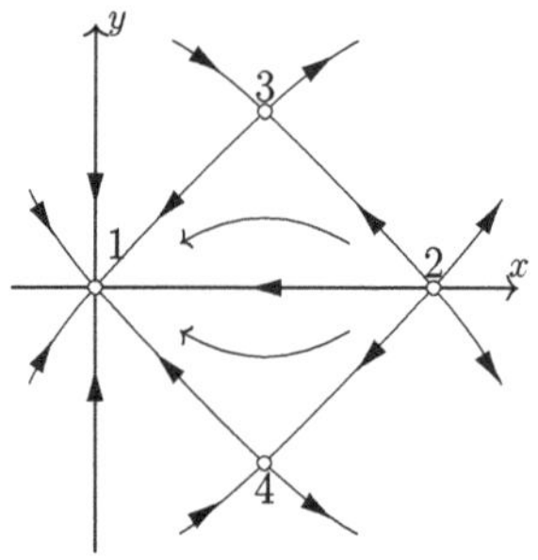

Figure 5.9: Phase portrait for the system in Problem 5.1. Fixed points: 1—stable node at (0, 0), 2—unstable node at (2, 0), 3 and 4—saddles at $(1, 2^{-1/2})$ and $(1, -2^{-1/2})$. As the eigenvalues of the Jacobian at the fixed points are non-negative, we can assume that the orbits do not flip and follow the sketched flows.

Solution to Problem 5.2.

(i) The Jacobian for the system is

$$J = \begin{pmatrix} b & 0 \\ 0 & 2 \end{pmatrix},$$

which has eigenvalues $\lambda_1 = b$ and $\lambda_2 = 2$. Since

$$|\det J| = 2b,$$

the system is area preserving when $b = 0.5$, in which case $|\det J| = 1$.

(ii) To obtain the image of the line $y = x$ after one iteration of the system, we set $y_0 = x_0$ with $x_0 \in [0, 1]$, which gives

$$y_1 = 2y_0 \bmod 1 = 2x_0 \bmod 1.$$

Hence $y_1 = 2x_0$ for $0 \le x_0 \le 0.5$, and $y_1 = 2x_0 - 1$ for $0.5 < x_0 < 1$.
Looking back at the system, we observe that

$$y_1 = 2x_0 = \frac{2}{b}x_1$$

when $x_1 \in [0, b/2]$ since $x_1 = bx_0$ for $x_0 \in [0, 1/2]$. Furthermore,

$$y_1 = 2x_0 - 1 = \frac{2}{b}x_1 - 1 - \frac{1}{b}$$

when $x_1 \in (\frac{b}{2} + \frac{1}{2}, b + \frac{1}{2}]$ since $x_1 = \frac{1}{2} + bx_0$ for $x_0 \in (1/2, 1]$. Thus, the image of the line $y = x$ is given by two straight segments shown in Figure 5.10.

(iii) The Jacobian matrix is diagonal and does not depend on n, so the Lyapunov exponents LE_1 and LE_2 in the x- and y-directions are given directly by

$$LE_1 = \lim_{n\to\infty} \frac{1}{n} \ln|\lambda_1^n| = \lim_{n\to\infty} \frac{1}{n} \ln b^n = \ln b,$$
$$LE_2 = \lim_{n\to\infty} \frac{1}{n} \ln|\lambda_2^n| = \lim_{n\to\infty} \frac{1}{n} \ln 2^n = \ln 2.$$

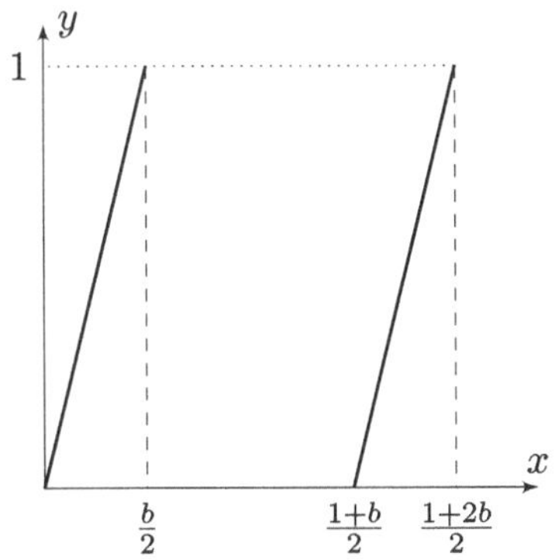

Figure 5.10: Image of the line $y = x$ after one iteration of the map in Problem 5.2. The image consists of two segments: the segment $y = \frac{2}{b}x$ for $x \in [0, \frac{b}{2}]$, which is the image of $y_0 = x_0$ for $x_0 \in [0, \frac{1}{2}]$, and the segment $y = \frac{2}{b}x - 1 - \frac{1}{b}$ for $x \in [\frac{1+b}{2}, \frac{1+2b}{2}]$, which is the image of $y_0 = x_0$ for $x_0 \in (\frac{1}{2}, 1]$.

Solution to Problem 5.3.

(i) To find the inverse map, we express x_n and y_n in terms of x_{n+1} and y_{n+1}. From the second equation, we obtain $x_n = y_{n+1}$, and from the first

$$y_n = x_{n+1} - 1 + ax_n^2 = x_{n+1} - 1 + ay_{n+1}^2.$$

Thus,

$$\begin{cases} x_n = y_{n+1}, \\ y_n = x_{n+1} - 1 + ay_{n+1}^2. \end{cases}$$

After inverting time, i. e., replacing n by $n + 1$ and $n + 1$ by n, we obtain the inverse map

$$\begin{cases} x_{n+1} = y_n, \\ y_{n+1} = x_n - 1 + ay_n^2. \end{cases}$$

(ii) Fixed points are the roots of the following algebraic system

$$\begin{cases} x = 1 + y - ax^2, \\ y = x. \end{cases}$$

They lie on the line $y = x$, with their coordinates determined by the quadratic equation obtained by substitution, namely,

$$x = 1 + x - ax^2 \quad \text{or} \quad 1 = ax^2.$$

Thus, we have two fixed points $X_{1,2} = Y_{1,2} = \pm\sqrt{a^{-1}}$, which exist when $a > 0$. Stability of the fixed points is determined by the eigenvalues of the Jacobian matrix

$$J = \begin{pmatrix} -2ax & 1 \\ 1 & 0 \end{pmatrix}.$$

The eigenvalues $\lambda_{1,2}$ are the roots of the characteristic equation

$$\lambda^2 + 2ax\,\lambda - 1 = 0,$$

namely,

$$\lambda_{1,2} = -ax \pm \sqrt{(ax)^2 + 1}.$$

For stability, we require $|\lambda_{1,2}| < 1$, which holds when the following two conditions are satisfied simultaneously:

$$\lambda_1 = -ax + \sqrt{(ax)^2 + 1} < 1,$$
$$\lambda_2 = -ax - \sqrt{(ax)^2 + 1} > -1.$$

The first inequality holds if $ax > 0$, and the second holds if $ax < 0$. This means that the eigenvalues cannot simultaneously lie between -1 and 1, and therefore neither of the two fixed points can be stable.
Indeed, for the fixed point $X_1 = Y_1 = \sqrt{a^{-1}}$ the eigenvalues are

$$\lambda_1 = -\sqrt{a} + \sqrt{a+1} < 1 \quad \text{(and positive)},$$
$$\lambda_2 = -\sqrt{a} - \sqrt{a+1} < -1,$$

and therefore this fixed point is a saddle.
For the fixed point $X_2 = Y_2 = -\sqrt{a^{-1}}$, the eigenvalues are

$$\lambda_1 = \sqrt{a} + \sqrt{a+1} > 1,$$
$$\lambda_2 = \sqrt{a} - \sqrt{a+1} > -1 \quad \text{(and negative)},$$

and this fixed point is also a saddle.
At $a = 0$ we have a tangent bifurcation of a very special kind: One eigenvalue satisfies $\lambda_1 = 1$, as required for a tangent bifurcation, but in addition the second eigenvalue is $\lambda_2 = -1$, indicating a flip bifurcation. This combination makes the bifurcation non-typical, and as a consequence both fixed points become saddles.

Solution to Problem 5.4.

(i) For the coordinates of the fixed points we have $y = (1 - B)x$, where x is defined by

$$x = (1 - B)x + 1 - ax^2,$$

which simplifies to

$$ax^2 + Bx - 1 = 0.$$

This yields

$$2aX = -B \pm \sqrt{B^2 + 4a}, \quad Y = (1 - B)X.$$

Hence, there are two fixed points, which exist only when $B^2 + 4a > 0$, i. e., when $a > -B^2/4$.
The Jacobian matrix associated with the system is

$$J = \begin{pmatrix} -2ax & 1 \\ 1 - B & 0 \end{pmatrix}.$$

The corresponding eigenvalues are obtained from

$$(-2ax - \lambda)(-\lambda) - 1(1 - B) = 0,$$

which simplifies to

$$\lambda^2 + 2ax\,\lambda + B - 1 = 0.$$

For asymptotic stability of the fixed point at (X, Y), the eigenvalues of the Jacobian at this point must satisfy

$$|\lambda_{1,2}| < 1.$$

The roots of the characteristic equation are

$$\lambda_{1,2} = -aX \pm \sqrt{(aX)^2 + 1 - B},$$

and therefore the conditions $|\lambda_{1,2}| < 1$ require

$$-aX - \sqrt{(aX)^2 + 1 - B} > -1,$$
$$-aX + \sqrt{(aX)^2 + 1 - B} < 1.$$

These inequalities hold simultaneously only when $|2aX| < B$. Indeed, from the first inequality we obtain

$$aX + 1 > \sqrt{(aX)^2 + 1 - B} \quad \Longrightarrow \quad (aX + 1)^2 > (aX)^2 + 1 - B,$$

from which we get $2aX > -B$. Since squaring introduces extraneous solutions, we must also impose the additional condition

$$aX + 1 > 0.$$

Similarly, from the second inequality we have

$$1 - aX > \sqrt{(aX)^2 + 1 - B} \quad \Longrightarrow \quad (1 - aX)^2 > (aX)^2 + 1 - B,$$

from which we get $2aX < B$. Again, to avoid extraneous solutions we must impose the additional condition

$$1 - aX > 0.$$

Certainly, (X_1, Y_1) does not satisfy the stability condition $|2aX| < B$ since

$$2aX_1 = -B - \sqrt{B^2 + 4a} < -B,$$

and is therefore unstable.
However, (X_2, Y_2) can satisfy the condition $|2aX| < B$ since

$$2aX_2 = -B + \sqrt{B^2 + 4a} < B$$

holds when

$$a < \frac{3}{4}B^2.$$

Since the fixed points exist only when $a > -B^2/4$, we obtain a stable fixed point exactly when

$$-\frac{1}{4}B^2 < a < \frac{3}{4}B^2.$$

We note that the condition $|2aX| < B$ is necessary but not sufficient for the stability of (X_2, Y_2). A sufficient condition requires the additional restriction

$$-1 < aX < 1.$$

When $a > \frac{3}{4}B^2$, both fixed points are unstable.
Consider the case $4a = 3B^2$, which gives

$$2aX_{1,2} = -B \pm \sqrt{B^2 + 3B^2} = -B \pm 2B.$$

As we have seen previously, (X_1, Y_1) is always unstable, meaning that at least one of the eigenvalues of the Jacobian at this fixed point lies outside the interval $[-1, 1]$. To examine this more closely, we compute both eigenvalues for

$$X_1 = -\frac{3B}{2a} \quad \text{when } a = \frac{3}{4}B^2 \text{ and } B = 0.5.$$

The characteristic equation in this case is

$$\lambda^2 - 1.5\lambda - 0.5 = 0,$$

which has solutions

$$\lambda_1 = \frac{-\frac{3}{2} + \sqrt{\frac{9}{4} + 2}}{2} = 1.78, \quad \lambda_2 = \frac{-\frac{3}{2} - \sqrt{\frac{9}{4} + 2}}{2} = -0.28.$$

Thus, X_1 has one stable direction (since $|\lambda_2| < 1$) and one unstable direction (since $|\lambda_1| > 1$), and is therefore a saddle point.
As for (X_2, Y_2), we have found that it is stable when $4a < 3B^2$ and unstable when $4a > 3B^2$. When $4a = 3B^2$, we expect that one of the eigenvalues of the Jacobian at this fixed point should be equal to 1 or −1, while the other lies in the interval $(-1, 1)$. We can verify this under the given conditions.
When $4a = 3B^2$ we have

$$X_2 = \frac{B}{2a},$$

and, taking $B = 0.5$, the characteristic equation becomes

$$\lambda^2 + 0.5\lambda - 0.5 = 0,$$

which has solutions

$$\lambda_1 = \frac{-\frac{1}{2} + \sqrt{\frac{1}{4} + 2}}{2} = 0.5, \quad \lambda_2 = \frac{-\frac{1}{2} - \sqrt{\frac{1}{4} + 2}}{2} = -1.$$

Altogether, as expected, X_2 possesses one stable direction (since $|\lambda_1| < 1$) and one neutral direction (since $|\lambda_2| = 1$), and is therefore a marginal fixed point.

(ii) Solutions X_3 and X_4 of

$$a^2x^2 - aBx + B^2 - a = 0,$$

satisfy

$$X_3 + X_4 = \frac{B}{a}, \quad X_3X_4 = \frac{B^2 - a}{a^2}.$$

The product of the Jacobian matrices associated with the 2-periodic solution is

$$J(X_3)J(X_4) = \begin{pmatrix} -2aX_3 & 1 \\ 1 - B & 0 \end{pmatrix} \begin{pmatrix} -2aX_4 & 1 \\ 1 - B & 0 \end{pmatrix} = \begin{pmatrix} 4a^2X_3X_4 + 1 - B & -2aX_3 \\ -2a(1 - B)X_4 & 1 - B \end{pmatrix}.$$

The associated eigenvalues are solutions of

$$\lambda^2 - (4a^2X_3X_4 + 2 - 2B)\lambda = -(1 - B)^2.$$

For $B = 1$, the characteristic equation reduces to

$$\lambda[\lambda - (4a^2X_3X_4)] = 0,$$

which implies that the eigenvalues are

$$\lambda_1 = 0, \quad \lambda_2 = 4a^2X_3X_4.$$

We deduce (from the formula for X_3X_4) that

$$\lambda_2 = 4a^2\frac{1-a}{a^2} = 4(1-a),$$

so that $|\lambda_2| < 1$ only when

$$\frac{3}{4} < a < \frac{5}{4},$$

which is precisely the interval of values of a for which a stable 2-cycle exists when $B = 1$.

Remark. When $B = 1$, the given map reduces to the one-dimensional iteration

$$x_{n+1} = 1 - ax_n^2,$$

a well-known quadratic map closely related to the logistic map. This map will be examined in the next chapter in connection with the Mandelbrot set.

6 Fractals and fractal dimensions

In the previous two chapters, we have encountered so-called *strange attractors* forming in 3D flows and 2D maps. These objects cause chaotic regimes in the corresponding dynamical systems and have a very specific geometry. In this chapter, we will focus on the study of such geometry, namely, on geometric structures that have non-integer dimensions and are known as *fractals*.

6.1 Maps on the complex plane

Consider the homogeneous linear map on the complex plane $z_{n+1} = Cz_n$, that is, the variable z and the parameter C are complex numbers ($z, C \in \mathbb{C}$). The solution to this map is $z_n = C^n z_0$, which can easily be seen from the chain $z_n = Cz_{n-1} = C^2 z_{n-2} = \cdots = C^n z_0$.

Now, using the exponential representation of C,

$$C = \mathbb{R}(C) + i\mathfrak{I}(C) = |C|\, e^{i\theta},$$

where $|C| = \sqrt{\mathbb{R}(C)^2 + \mathfrak{I}(C)^2}$ is the modulus and $\theta = \arctan(\mathfrak{I}(C)/\mathbb{R}(C))$ is the argument of the complex number C, we have

$$z_n = C^n z_0 = z_0\, |C|^n e^{i\theta n} = z_0\, |C|^n (\cos(n\theta) + i\sin(n\theta)).$$

Thus, $z_n \rightarrow 0$ if $|C| < 1$. All of this can be stated in the following way: The map $z_{n+1} = Cz_n$ has a fixed point $z = 0$ which is stable if $|C| < 1$. Furthermore, we can state that if $|C| > 1$, all orbits (except for the one with $z_0 = 0$) are unbounded, while if $|C| < 1$, all orbits are bounded.

One of the central problems in relation to maps on the complex plane is whether orbits are bounded or not. Let us consider the autonomous map on the complex plane $z_{n+1} = f(z, C)$. For a given value of the parameter C, we find that some orbits are bounded and some are not. Thus, our first task is to determine whether, for a given value of C, there exist bounded solutions. If bounded solutions exist, we can further examine them and check whether they converge to fixed or periodic solutions, or whether they are chaotic.

Definition 6.1. Consider the quadratic map on the complex plane

$$z_{n+1} = z_n^2 + C, \tag{6.1}$$

where $C, z_n \in \mathbb{C}$. The **Mandelbrot set** is the set of C values for which there exist bounded solutions (i. e., there is at least one value z_0 such that $|z_n| \nrightarrow \infty$).

Obviously, bounded orbits exist if there is a stable fixed point. Thus, values of C for which there is a stable fixed point belong to the Mandelbrot set. Let us find the fixed

https://doi.org/10.1515/9783111348339-006

points of the complex map (6.1) and analyze their stability. As we know, these fixed points are the roots of the algebraic equation

$$z = z^2 + C, \quad \text{or} \quad z^2 - z + C = 0.$$

The solutions of this quadratic equation are given by

$$Z_{1,2} = \frac{1 \pm \sqrt{1-4C}}{2}. \tag{6.2}$$

As we know, for maps defined on the real axis, $u_{n+1} = f(u_n)$, the fixed point U is stable if $|f'(U)| < 1$. Similarly, for the map on the complex plane $z_{n+1} = f(z_n)$, a fixed point Z is stable if $|f'(Z)| < 1$: In this case $f'(Z)$ is a complex number and we consider its modulus. To analyze the stability of the fixed points, we should compute the derivative of the function

$$f(z) = z^2 + C,$$

i. e.,

$$f'(z) = 2z.$$

So, for the first fixed point given in (6.2),

$$f'\left(Z_1 = \frac{1 + \sqrt{1-4C}}{2}\right) = 1 + \sqrt{1-4C},$$

where C is a complex number. Let

$$\sqrt{1-4C} = \mathbb{R} + i\mathbb{J},$$

and by the definition of the square root, we have $\mathbb{R} \geq 0$. Furthermore,

$$|f'(Z_1)| = \sqrt{(1+\mathbb{R})^2 + \mathbb{J}^2},$$

and since $(1+\mathbb{R}) > 1$, the modulus of the derivative at Z_1 is always greater than 1. Therefore, $Z_1 = \frac{1+\sqrt{1-4C}}{2}$ is unstable for any value of C.

At the second fixed point given in (6.2), the derivative is

$$f'\left(Z_2 = \frac{1 - \sqrt{1-4C}}{2}\right) = 1 - \sqrt{1-4C}.$$

For Z_2 to be stable, we need $|f'(Z_2)| < 1$, and this now becomes

$$|f'(Z_2)| = \sqrt{(1-\mathbb{R})^2 + \mathbb{J}^2} < 1, \tag{6.3}$$

a condition that is satisfied for some values of C.

Definition 6.2. A set of values of the parameter C in map (6.1) (in the complex plane $\mathbb{C}$) for which Z_2 is stable (that is, $(1-\mathbb{R})^2+\mathbb{J}^2<1$, where $\mathbb{R}$ and $\mathbb{J}$ are the real and imaginary parts of $\sqrt{1-4C}$), is called the **cardioid**. The cardioid is a part of the Mandelbrot set, as shown in Figure 6.1.

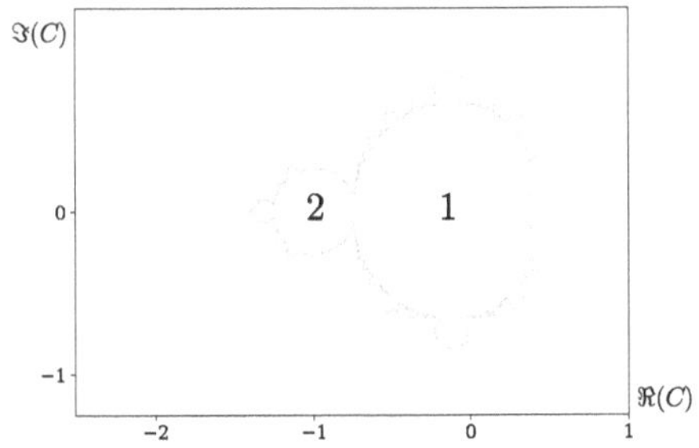

Figure 6.1: Image of the Mandelbrot set. It consists of regions in the complex plane corresponding to the multiple stable periodic solutions of (6.1). The region marked '1' is known as the cardioid and is defined by the inequality (6.3). This region contains all values of the parameter C for which the fixed point (i. e., the one-periodic solution) is stable. The region marked '2', which is a circle given by (6.4), corresponds to the existence of stable two-periodic solutions.

The Mandelbrot set also includes values of C for which the map (6.1) has stable periodic solutions. For example, 2-periodic solutions are composed of the two fixed points of the second iterate of the map, namely,

$$z_{n+1}=z_n^2+C \quad \Rightarrow \quad z_{n+2}=z_{n+1}^2+C=\left(z_n^2+C\right)^2+C=z_n^4+2Cz_n^2+C^2+C.$$

Thus, the 2-periodic solutions are roots of the algebraic equation

$$z=z^4+2Cz^2+C^2+C.$$

This is a fourth-order polynomial, which can be factorized in the following way:

$$(z^2-z+C)(z^2+z+C+1)=0,$$

with the periodic solution given by the two roots of the quadratic polynomial in the second bracket. From Vieta's theorem, it follows that $Z_3Z_4=C+1$.

To analyze the stability of the two-periodic solution, we use the following technique developed for maps with real variables (see Chapter 2):

$$\left.\frac{df^{(2)}(z)}{dz}\right|_{z=Z_3}=f'(Z_3)f'(Z_4)=(2Z_3)(2Z_4)=4Z_3Z_4=4(C+1).$$

So, the 2-periodic solution (Z_3,Z_4) is stable if $|4(C+1)|<1$ or

$$|C+1|<1/4, \tag{6.4}$$

which is a circle of radius 1/4 with the center at $(-1,0)$. The values of C for which the two-periodic solution is stable cover the area marked as '2' in Figure 6.1.

Furthermore, we can find all values of the parameter C for which the map (6.1) has stable n-periodic solutions for $n = 3, 4, \ldots \infty$. For each value of n, we identify a set of C values that contributes an additional region in the complex plane to the Mandelbrot set shown in Figure 6.1. The collection of all these regions forms a complex geometric object that has a finite area but an infinite perimeter. Such objects are called fractals, and thus the Mandelbrot set has a fractal structure.

6.2 Monster curves

Fractals are geometric objects which have many interesting features, including self-similarity, which means that they are composed of parts similar to the whole object. Also, the description of fractals involves the introduction of non-integer dimensions. The first fractals were introduced in the early 20th century as lines of infinite length (so-called *monster curves*).

From calculus, we know that if the curve can be described by a differentiable function $y = f(x)$, with $x \in [a, b]$, then its length can be found analytically. To find it, we divide the curve into small pieces and replace these pieces with segments of straight line connecting their end points (see Figure 6.2), that is,

$$L = \sum_{i=1}^{n} L_i,$$

where $L_i \sim \sqrt{\Delta x_i^2 + \Delta y_i^2}$. At the limit of infinitesimal division, the length of the curve converges to the value given by the definite integral following:

$$L = \int_{x=a}^{x=b} \sqrt{dx^2 + dy^2} = \int_{x=a}^{x=b} \sqrt{1 + \left(\frac{dy}{dx}\right)^2} dx = \int_a^b \sqrt{1 + (f'(x))^2} dx.$$

If $f'(x)$ is defined everywhere and $f'(x) \neq \infty$, then L is found and finite. This technique for finding the length of curve does not work with certain curves, namely, the curves that can't be described by differentiable functions (which are called fractals or monster curves). Next are a few examples of such curves.

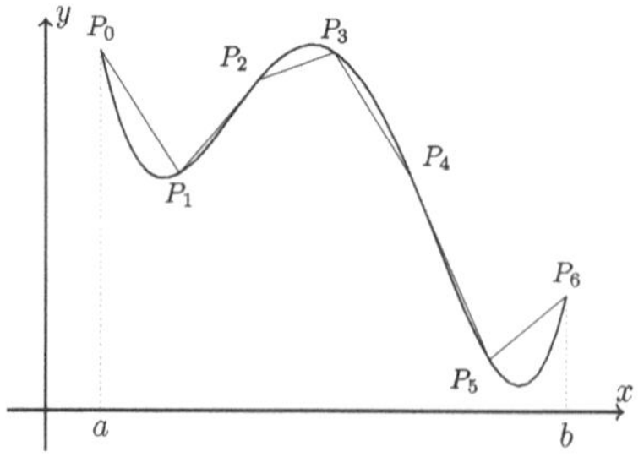

Figure 6.2: On finding the length of a curve given by differentiable function $f(x) \in C^1, x \in [a, b]$.

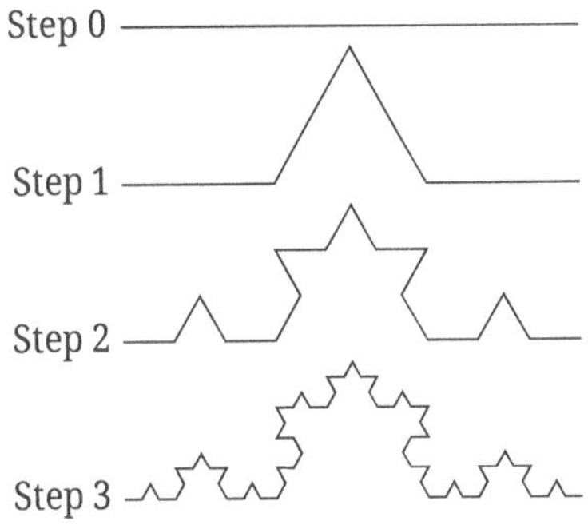

Figure 6.3: First three steps in building the Koch curve.

Example 6.1 (Koch curve). This curve was introduced by the Swedish mathematician Helge von Koch in 1904. The construction of the Koch curve takes an infinite number of steps, of which the first three are illustrated in Figure 6.3.

- **Step 0:** Start with a segment of length 1.
- **Step 1:** Divide the segment into three equal parts and replace the one in the middle with two new segments as shown in Figure 6.3. Now (at Step 1), our object contains four segments, each of length 1/3, and the length of the curve is 4/3.
- **Step 2:** Repeat the procedure given in Step 1, that is, replace the middle third of each of the four segments with two new segments. Now (at Step 2), our object contains 16 segments, each of length 1/9, and the length of the curve is 16/9.
- **Step n:** By repeating the procedure n times, we obtain a curve containing 4^n segments, each of length $1/3^n$. The length of the curve is $L = 4^n/3^n$, which tends to infinity as $n \to \infty$.

Example 6.2 (Koch snowflake). The object known as the Koch snowflake is obtained when we begin with an equilateral triangle and follow the procedure, replacing the middle third of each segment with a “$\wedge$”, and repeat this infinitely. The resulting geometric object indeed resembles a snowflake and is illustrated in Figure 6.4.

Figure 6.4: Koch snowflake.

6.3 Fractal dimensions

As an introduction to fractal dimension, we recall that a point in space has no size and its dimension is zero; a line has length but does not have an area and its dimension is 1; a surface has an area but not a volume and its dimension is 2. The corresponding illustration is given in Figure 6.5.

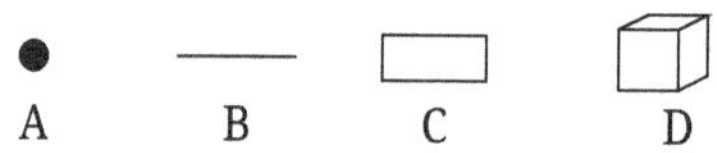

Figure 6.5: An illustration of 0-, 1-, 2- and 3-dimensional objects. A: A point has zero length and its dimension is 0. **B:** A line has length but not area and has a dimension of 1. **C:** A square or rectangle has an area (but no volume) and has a dimension of 2. **D:** A cube has a volume and is a 3-dimensional object.

Length measurement can be formalized in the following way. Use a ruler of a certain size: If a ruler of size ε is applied N times to cover the curve, then the length can be estimated as $L \approx N\varepsilon$. When we reduce the length of the ruler, ε, the number of its applications, N, required to cover the curve increases. If the product $N\varepsilon$ converges as $\varepsilon \to 0$, we consider its limit as the length of the curve. However, this convergence is not always observed. A famous example is the length of a coastline, i. e., the British coastline [16]. A ruler of size 200 km gives 2,400 km for the length of the British coastline. A ruler of size 50 km gives 3400 km for this coastline. The length increases further with smaller rulers, as finer details of the coastline contribute to the total length when the ruler size is reduced.

The same procedure can be applied to area measurement. For a surface (let us say a 1×1 square),

(1) A 1×1 square (as a measure) covers the original shape.
(2) A $1/10 \times 1/10$ square (as a measure) has to be applied 100 times to cover the shape completely (illustrated in Figure 6.6).

The question is: How many scaled squares make up the original square? The answer is that if ε is the side length of the *measuring square*, then the number of measuring squares covering the original is $N = 1/\varepsilon^2$, provided that ε is small enough, $\varepsilon \ll 1$.

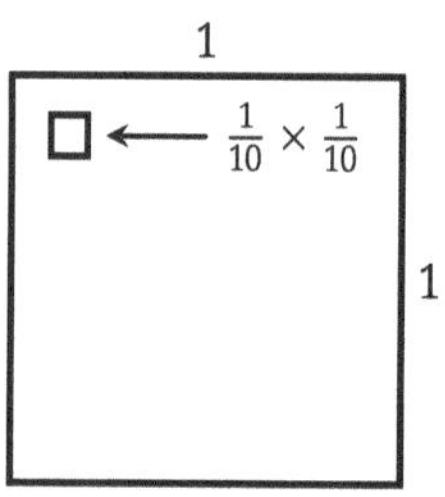

Figure 6.6: A square of 1×1 can be covered by 100 smaller squares of size $\varepsilon = 1/10$.

To introduce a measure of fractal dimension known as the *capacity dimension*, we note that for a curve of finite length,

$$L \to N(\varepsilon)\varepsilon \quad \Rightarrow \quad N(\varepsilon) \to \frac{L}{\varepsilon}, \quad \text{as } \varepsilon \to 0.$$

Similarly, for a surface of finite area,

$$A \to N(\varepsilon)\varepsilon^2 \quad \Rightarrow \quad N(\varepsilon) \to \frac{A}{\varepsilon^2}, \quad \text{as } \varepsilon \to 0.$$

Similarly, for the volume of an object in space,

$$V \to N(\varepsilon)\varepsilon^3 \quad \Rightarrow \quad N(\varepsilon) \to \frac{V}{\varepsilon^3}, \quad \text{as } \varepsilon \to 0.$$

The dimension of a line is 1, the dimension of a surface is 2, and the dimension of a volume is 3. Furthermore, for a point, a measuring object of any size gives $N = 1$, which can be written as ε^{-0}. The geometric dimension of a point is 0. As we see, the geometric dimension is given by the parameter d in the formula defining the number of *measuring boxes* (which, depending on the geometric object we deal with, may be represented by a line segment, square, or cube) required to cover the object, as follows:

$$N(\varepsilon) \to \frac{L}{\varepsilon^d}, \quad \text{as } \varepsilon \to 0.$$

Following this observation, we can define the dimension of a fractal object through the following procedure. Take measuring boxes of size ε and cover the object with these boxes. Count the number of boxes required to cover the object. If there exists a number d_c such that

$$N(\varepsilon) \to C\,\varepsilon^{-d_c} \quad \text{as } \varepsilon \to 0,$$

then d_c is used to define the dimension of the object. This quantity is called the *capacity dimension* or *box-counting dimension*. Taking logarithms of both sides of the expression, we obtain

$$\ln N(\varepsilon) \to \ln C + d_c \ln\left(\frac{1}{\varepsilon}\right) \quad \text{as } \varepsilon \to 0.$$

From this relation, we can express the capacity dimension d_c as

$$d_c = \lim_{\varepsilon \to 0} \frac{\ln N(\varepsilon) - \ln C}{\ln(\frac{1}{\varepsilon})} = \lim_{\varepsilon \to 0} \frac{\ln N(\varepsilon)}{\ln(\frac{1}{\varepsilon})} \tag{6.5}$$

since $\ln C$ is a constant, while $\ln N(\varepsilon) \to \infty$ as $\varepsilon \to 0$.

This definition has certain drawbacks and should be used carefully. The main mathematical issue arises from the fact that, for different sequences $\{\varepsilon_i\}$ such that $\varepsilon_i \to 0$ as $i \to \infty$, there may exist different limits, meaning that $\lim_{\varepsilon\to 0}$ does not exist at all. This limitation can be overcome for a class of self-similar fractals by using the sequence $\varepsilon_i = s^i$, where s is the self-similarity scaling of the fractal object under consideration. This ensures that the part of the object covered by a measuring box is a smaller copy of the entire object.

We now consider a few examples of fractal objects for which we will compute their capacity dimensions.

6.4 Cantor set

The Cantor set is one of the best-known objects with a fractal structure. It was discovered in the 18th century. There are many modifications of this object, the most classical being the so-called middle-third version, which is constructed according to the following procedure:

- **Step 0:** Start with a straight line segment of length 1.
- **Step 1:** Divide this segment into three equal pieces and remove the middle one, leaving the remaining two smaller segments.
- **Step 2:** Divide each of the two segments obtained in the previous step into three equal pieces and remove the middle one.
- **Further steps:** Repeat this procedure infinitely many times, removing the middle third interval from each subsequent segment.

The Cantor set is the collection of all points that remain after this infinite process, which is illustrated in Figure 6.7.

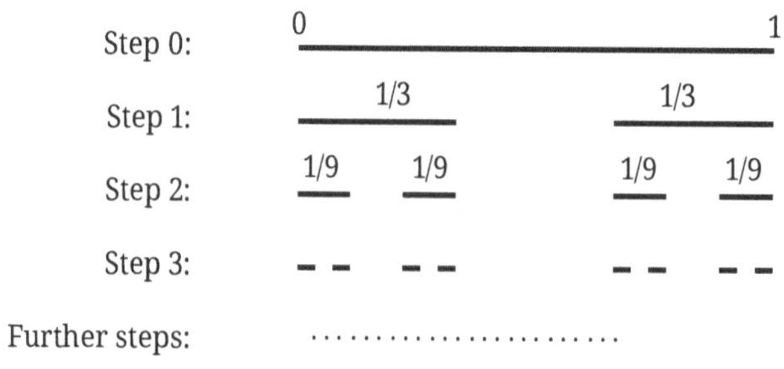

Figure 6.7: Illustration of the construction of the Cantor middle-third set.

To measure the capacity dimension of the obtained Cantor set, we note that it is self-similar and the scale of self-similarity is 1/3. The two sub-objects (segments) obtained in

Step 1 are similar to the original object (at Step 0) but are three times smaller. The subobjects generated in Step 2 are three times smaller than those obtained in Step 1, and so on. At Step n we obtain 2^n objects (segments) of size $1/3^n$.

We use a ruler as a measuring box for Cantor set. We note that the length of the object at Step 0 is 1, $L = 1$, and a ruler of size 1 covers the object in one go. At Step 1, we can use a ruler of size 1/3 to find that it takes two measuring objects to cover the set. Generally, at Step n we need 2^n measuring boxes of size $1/3^n$ to cover the entire object. These measurements are summarized in the Table 6.1.

Table 6.1: Measurements for determining the capacity dimension of the Cantor middle-third set.

Step	Scale	Number of boxes
0	1	1
1	1/3	2
2	$1/3^2 = 1/9$	4
3	1/27	8
⋮	⋮	⋮
n	$1/3^n$	2^n

The procedure we use to measure the capacity dimension can also be viewed as follows. We apply the ruler not to the intermediate objects but to the final object representing the Cantor set. A measuring box of size 1 covers all points forming the Cantor set. Two measuring boxes of size 1/3 cover the entire Cantor set, and the parts covered by each box are smaller copies of the whole set. We then successively reduce the size of the measuring box by a factor of three and count the number of boxes required to cover the entire Cantor set.

Using the number of measuring boxes required to cover the object for a given box size, we find

$$d_c = \lim_{\varepsilon\to 0} \frac{\ln(N(\varepsilon))}{\ln(\frac{1}{\varepsilon})} = \frac{\ln(2^n)}{\ln(3^n)} = \frac{n\ln(2)}{n\ln(3)} = \frac{\ln(2)}{\ln(3)} \approx 0.53.$$

So, the fractal dimension of the Cantor set is a fractional number between 0 (the dimension of a point) and 1 (the dimension of a line). This indicates that the length of the Cantor set is zero. Indeed, we start with a segment of length 1. On the first step, we obtain two segments with a total size of 2/3. On the n-th step, we have 2^n segments of size 1/3 each. The total size of all segments $(2/3)^n$, which tends to zero as n tends to infinity.

Example 6.3. Cantor's 'middle fifth' is produced by the iterated process of dividing each interval into five segments and removing the open middle (third) segment of the previous segments, starting with the unit interval. Find the capacity dimension of this Cantor set.

Solution:
To find the capacity dimension of Cantor's middle fifth, we follow the scaling factor $\varepsilon = 1/5$ and count the number of non-empty segments at each step of the construction:

$$\begin{array}{rlccc} \text{Step} & 1 & 2 & \cdots & n \\ \text{Size of segments, } \varepsilon & 1/5 & (1/5)^2 & \cdots & (1/5)^n \\ \text{Number of segments, } N(\varepsilon) & 4 & 4^2 & \cdots & 4^n. \end{array}$$

Hence, for the capacity dimension of Cantor's middle fifth, we have

$$d_c = \lim_{\varepsilon \to 0} \frac{\ln N(\varepsilon)}{\ln(1/\varepsilon)} = \frac{\ln 4^n}{\ln 5^n} = \frac{\ln 4}{\ln 5} \approx 0.8614.$$

Remark. It is important that we follow the self-scaling factor $\varepsilon = 1/5$ of the constructed fractal. Note that two segments of size 2ε can cover the object obtained in Step 1, and one could continue using measuring boxes of size $\varepsilon = (2/5)^n$, which would give a different (and incorrect) capacity dimension.

The Cantor set is an infinite set of points, which is why it is often referred to as *Cantor dust*. Looking at the construction procedure for Cantor's middle third, we note that the precursor of the Cantor set at Step 0 is the full interval $[0, 1]$, consisting of all points whose coordinates can be written as

$$0.a_1 a_2 a_3 \ldots a_n \ldots,$$

where each digit a_i may be any element of $\{0, 1, 2\}$ in the ternary (base-3) system.

At Step 1, all points in the interval (1/3, 2/3) are removed. In ternary notation, this corresponds to removing all points whose first digit is $a_1 = 1$. Thus, the points that remain after Step 1 are precisely those for which a_1 is either 0 or 2.

At Step 2, we remove the middle third from each of the two remaining segments. In ternary terms, this eliminates all points whose second digit is $a_2 = 1$. The points that remain are those for which a_2 is either 0 or 2.

Continuing this process indefinitely, at each step we remove all points whose ternary expansion contains a digit 1 in the corresponding position. After infinitely many steps, we obtain the Cantor set S, consisting of all points

$$x = 0.a_1 a_2 a_3 \ldots a_n \ldots$$

in base 3 such that

$$a_i \neq 1 \quad \text{for all } i.$$

Thus, the Cantor set is precisely the set of numbers in $[0, 1]$ whose ternary expansions use only the digits 0 and 2.

6.5 Sierpiński triangle

Another classical example of a fractal object is given by the so-called Sierpiński triangle, which is named after the Polish mathematician Waclaw Sierpiński. The Sierpiński triangle is constructed using the following procedure:

- **Step 0:** Begin with an equilateral triangle with a side given by a unit segment (shown in Figure 6.8A).
- **Step 1:** (a) Find the midpoint of each of the three sides of the triangle. (b) Draw lines connecting these midpoints to form a new, central, inverted triangle. (c) Erase the central triangle, leaving three smaller triangles in the corners (the resulting object is shown in Figure 6.8B).

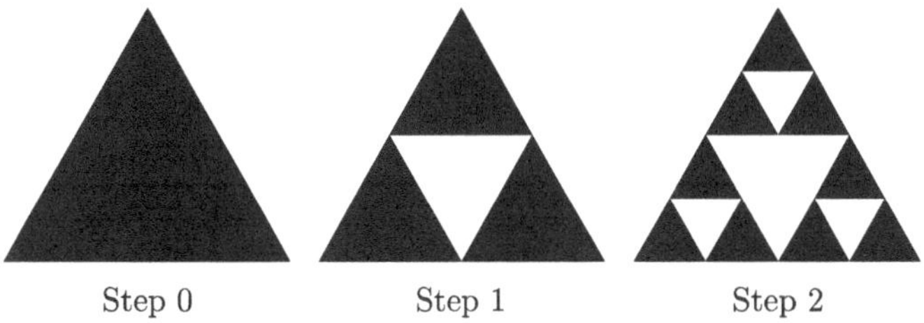

Figure 6.8: Illustration of the first three steps in the construction of the Sierpiński triangle.

- **Step 2:** Repeat the procedure described in Step 1 for each of the remaining corner triangles obtained in Step 1 (the resulting object is shown in Figure 6.8C).
- **Further steps:** Continue applying the procedure described in Step 1 to each of the triangles remaining after the previous step.

To calculate the capacity dimension of the Sierpiński triangle, we note that the self-similarity scale is 1/2, and as measuring objects, we shall use triangles of size $\varepsilon = (1/2)^n$ with $n \in \mathbb{N}_0$. The number of boxes (triangles) covering the Sierpiński triangle for $\varepsilon_i = (1/2)^i$, $i = 0, 1, 2, \ldots, n$, is given in Table 6.2:

Table 6.2: Measurements for determining the capacity dimension of the Sierpiński triangle.

Step	Scale	Number of boxes
0	1	1
1	1/2	6
2	1/4	9
⋮	⋮	⋮
n	$1/2^n$	3^n

Based on the presented data we find the capacity dimension of Sierpiński triangle as follows:

$$d_c = \lim_{\varepsilon \to 0} \frac{\ln(N)}{\ln(\frac{1}{2^n})^{-1}} = \frac{\ln(3^n)}{\ln(2^n)} = \frac{\ln 3}{\ln 2} \approx 1.6.$$

We note that for the capacity dimension of Sierpiński triangle is between 1 and 2, which tells us that the object is "more than a line" and "less than a surface". Indeed, it is easy to see that the area of the Sierpiński triangle is zero, while the length of its perimeter is infinite.

The area of the objects formed at various steps of the Sierpiński triangle construction is as follows:

- At Step 0: $A_0 = \frac{\sqrt{3}}{2}$.
- At Step 1: $A_1 = (\frac{3}{4})A_0$.
- At Step 2: $A_2 = (\frac{3}{4})A_1 = (\frac{3}{4})^2 A_0$.
- $\vdots$
- At Step n: $A_n = (\frac{3}{4})^n A_0$.
- $\vdots$

Looking at this sequence, we conclude that $A_n \to 0$ as $n \to \infty$.

Similarly, calculations of the Sierpiński triangle *perimeter* give

- At Step 0: We have one triangle of size 1, which has perimeter $L_0 = 3$.
- At Step 1: We have three triangles of size 1/2, each having perimeter 3/2, giving the total perimeter $L_1 = 3(3/2)$.
- At Step 2: We have nine triangles of size 1/4, each having perimeter 3/4, giving the total perimeter $L_2 = 9(3/4) = 3(3/2)^2$.
- $\vdots$
- At Step n: We have 3^n triangles of size $1/2^n$, with total perimeter

$$L_n = 3\left(\frac{3}{2}\right)^n.$$

Looking at the formula for L_n, we conclude that $L_n \to \infty$ as $n \to \infty$.

General note concerning fractal dimensions (supported by the calculations just shown for the Cantor set and the Sierpiński triangle),

- If $0 < d_c < 1$, then the length of the fractal object is zero.
- If $1 < d_c < 2$, then the length of the fractal object is infinite and the area of the object is zero.
- If $2 < d_c < 3$, then the area of the fractal object is infinite and the volume is zero.

Many other fractal objects can be constructed using slightly modified procedures based on the construction of the Sierpiński triangle. A fractal object, which is called the *Sierpiński square* or *Sierpiński carpet*, is constructed in the following way:

Figure 6.9: Illustration of the first three steps in the construction of the Sierpiński square.

- **Step 0:** Begin with a unit square (shown in Figure 6.9A).
- **Step 1:** Divide it into a 3×3 grid (nine equal smaller squares) and remove the middle square (the resulting object is shown in Figure 6.9B).
- **Step 2:** Take each of the eight remaining squares, divide each into nine smaller squares, and remove the middle one from each (the resulting object is shown in Figure 6.9C).
- **Further steps:** Continue this process of subdivision and removal for all squares obtained in the previous step.

To measure the capacity dimension of the Sierpiński square, we use a square as the measuring box and take into account the self-similarity scale, which is 1/3. The number of boxes covering the fractal object for various sizes of measuring boxes is given in Table 6.3.

Table 6.3: Measurements for determining the capacity dimension of the Sierpiński square.

Step	Scale	Number of squares
0	1	1
1	1/3	8
2	$1/3^2 = 1/9$	$(8 \times 9) - (8 \times 1) = 64$
⋮	⋮	⋮
n	$1/3^n$	8^n

Based on the obtained number of boxes, we calculate

$$d_c = \lim_{\varepsilon\to 0} \frac{\ln(N(\varepsilon))}{\ln(\frac{1}{\varepsilon})} = \frac{\ln(8)^n}{\ln(3)^n} = \frac{\ln(8)}{\ln(3)} \approx 1.893.$$

The capacity dimension of the Sierpiński square is again a fractional number between 1 and 2, indicating that the length of its perimeter is infinite, while its area is zero.

6.6 Fractals and chaos

It was stated in Chapter 5 that the chaotic regime in the Hénon map is associated with the emergence of a strange attractor, and this attractor has a fractal structure. Here, we will illustrate this point using the chaotic regime observed in the logistic map. It appears that the attractor formed through the period-doubling cascade in the logistic map at the value of parameter when chaos emerges is a fractal object, namely, it is a Cantor set.

As we seen in Chapter 2, the distance between neighboring points forming stable periodic solutions in one-dimensional maps decreases during the period-doubling cascade, following a geometric progression with a common ratio given by the second Feigenbaum constant, $\alpha \approx 2.5029$. To describe the fractal object associated with the emergence of chaos in the logistic map, we outline the period-doubling cascade in the following way.

As we remember, in the logistic map at $r = 3$, we have a flip bifurcation at the fixed point $U_2 = 2/3$, and this bifurcation leads to the formation of a stable period-2 oscillation. As the parameter r increases from 3 to $1 + \sqrt{6}$, the distance between the two points forming the stable two-periodic solution grows from 0 to

$$d_0 = \frac{\sqrt{3}}{2} \approx 0.866.$$

Each of these two points gives rise to two further points forming a four-periodic solution, and at the moment when this solution becomes unstable the distance between the points in each pair is approximately d_0/α, where α is the second Feigenbaum constant. Similarly, the maximal distance between points in each subsequent pair forming the eight-periodic solution is d_0/α^2, and so on. This iterative refinement can be used to construct a fractal object formed by the stable $2n$-periodic solutions as $n \to \infty$.

- **Step 0:** Begin with a segment representing all possible values of the roots forming the stable 2-periodic orbit. The size of this segment is d_0 (see Figure 6.10).
- **Step 1:** Plot two segments, which represent all possible values of the four roots forming the 4-periodic solution. Sizes of these segments can be approximated as d_0/α (Figure 6.10).
- **Step 2:** Plot four segments of size d_0/α^2, which represent all possible values of the eight roots forming the 8-periodic solution (shown in Figure 6.10).
- **Further steps:** Continue this process, plotting segments whose sizes are determined by the second Feigenbaum constant, thereby capturing all possible values of the roots forming the stable 2^n-periodic solutions.

This construction process, illustrated in Figure 6.10, indicates that the attractor in the logistic map at the value of the parameter r at which chaos emerges is a Cantor set.

Using the scaling factor α and a ruler as a measuring box, we find that at Step n, we have 2^n segments of size d_0/α^n (see Table 6.4). Based on this, we find that the capacity dimension of the attractor in the logistic map is

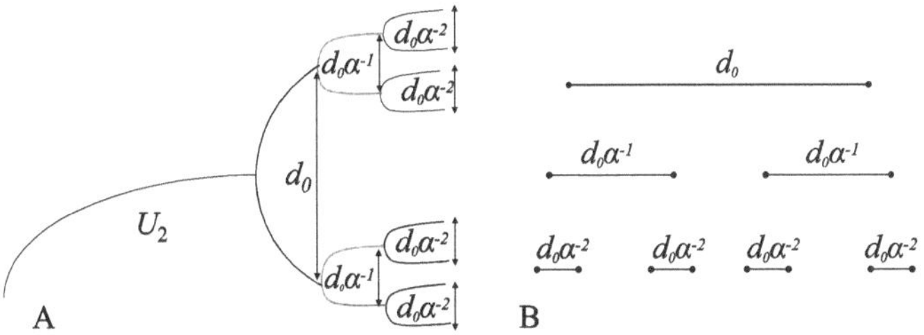

Figure 6.10: Self-similar structure of the period-doubling cascade. The period-doubling cascade exhibits a self-similarity scale determined by the second Feigenbaum constant $\alpha \approx 2.50$, manifested in the fractal arrangement of stable periodic solutions. **Panel A:** Schematic representation of the period-doubling cascade that arises as the parameter r in the logistic map increases from $r = 1$ to $r \approx 3.57$ (see Figure 2.7). **Panel B:** Corresponding ranges of root values forming 2-, 4-, and 8-periodic orbits, illustrated through the first three steps in the construction of a Cantor-like set.

$$d_c = \lim_{\varepsilon\to 0} \frac{\ln(N(\varepsilon))}{\ln(\frac{1}{\varepsilon})} = \frac{\ln(2^n)}{\ln(\alpha^n)} = \frac{n\ln 2}{n\ln\alpha} \approx \frac{\ln 2}{\ln 2.5029} \approx 0.756.$$

Thus, the attractor is a fractal object whose capacity dimension is less than 1; in fact, it is a Cantor set.

Table 6.4: Measurements for determining the capacity dimension of the attractor in the logistic map.

Attractor periodicity	Scale	Number of boxes
2	1	1
4	$1/\alpha$	2
8	$1/\alpha^2$	4
⋮	⋮	⋮
2^{n+2}	$(1/\alpha)^n$	2^n

6.7 Using maps to build fractals

There are many ways to use maps to construct fractal objects, and in what follows we consider several illustrative examples.

6.7.1 Designing a map to build the Cantor set

Consider the map

$$x_{n+1} = 3x_n \bmod 1, \quad x_0 \in [0,1],$$

which is a modification of the doubling map discussed in Chapter 2. The corresponding map-generating function can be written as

$$f(x) = \begin{cases} 3x, & \text{if } 0 \le x < \frac{1}{3}, \\ 3x - 1, & \text{if } \frac{1}{3} \le x < \frac{2}{3}, \\ 3x - 2, & \text{if } \frac{2}{3} \le x \le 1. \end{cases}$$

It has three branches and can be used to replicate the construction of the Cantor middle-third set. The removal of the middle third is modelled by modifying $f(x)$ so that $f(x) = 0$ for $1/3 < x < 2/3$. Using this approach, we construct a map that assigns the value zero to the middle third of every nonzero segment obtained in the previous iteration of the map, thereby generating the following Cantor set:

$$x_{n+1} = \begin{cases} 3x_n, & \text{if } 0 \le x_n < \frac{1}{3}, \\ 0, & \text{if } \frac{1}{3} \le x_n < \frac{2}{3}, \\ 3x_n - 2, & \text{if } \frac{2}{3} \le x_n \le 1. \end{cases}$$

This map can also be presented in a more compact way as follows:

$$x_{n+1} = \begin{cases} 0, & \text{if } 1/3 \le x_n < 2/3, \\ 3x_n \pmod 1, & \text{otherwise.} \end{cases}$$

The set of $x_0's$, such that $x_n \nrightarrow 0$ in this map is a Cantor set.

6.7.2 Combining maps to build the Cantor set

Another way to construct the Cantor set is by using a combination of maps. Consider the following two maps:

- $w_1 : x_{n+1} = \frac{1}{3}x_n,\ x_0 \in [0,1]$.
- $w_2 : x_{n+1} = \frac{1}{3}x_n + \frac{2}{3},\ x_0 \in [0,1]$.

Iterative use of the maps w_1 and w_2 leads to the construction of the Cantor set. In the first iteration, applying both maps to the initial segment $x_0 \in [0,1]$ produces two smaller segments,

$$x_1 \in \left[0, \frac{1}{3}\right] \quad \text{and} \quad x_1 \in \left[\frac{2}{3}, 1\right],$$

which corresponds to removing the middle third of the unit interval.

In the second iteration, we apply both maps to every point in the two remaining segments. This produces four smaller segments, each obtained by removing the middle third from the previous ones.

Continuing this process, each iteration removes the middle third from every surviving segment. The set of all possible values of x_n as $n \to \infty$ is precisely the Cantor middle-third set.

6.7.3 Combining maps to build the Koch curve

To generate the Koch curve, we combine four maps—one for each of the four segments that replace a single segment at every iteration (see Figure 6.3). Unlike the Cantor set, the Koch curve does not lie on a straight line so we require two-dimensional maps of the form

$$w_i : \begin{pmatrix} x_{n+1} \\ y_{n+1} \end{pmatrix} = A \begin{pmatrix} x_n \\ y_n \end{pmatrix} + \mathbf{b},$$

where each map w_i is defined by a linear transformation A (rotation and scaling) together with a translation vector $\mathbf{b}$. These four affine maps collectively generate the Koch curve through iterative refinement.

To obtain the first segment, we observe that it is simply the original segment contracted by a factor of three. This segment is therefore generated by applying the map w_1, which scales the original segment by 1/3,

$$w_1 : \begin{pmatrix} x_{n+1} \\ y_{n+1} \end{pmatrix} = \frac{1}{3} \begin{pmatrix} 1 & 0 \\ 0 & 1 \end{pmatrix} \begin{pmatrix} x_n \\ y_n \end{pmatrix}.$$

The image of the segment connecting the points $(0, 0)$ and $(1, 0)$ under this map is the segment joining

$$(0, 0) \quad \text{and} \quad \left(\frac{1}{3}, 0\right).$$

The second of the four segments obtained in the first step of constructing the Koch curve (see Figure 6.11) is also a copy of the original segment scaled by a factor of 1/3, but in addition it is shifted and rotated. Thus, the map w_2, like w_1, contracts the original segment by 1/3, then rotates it by $\pi/3$ anticlockwise, and finally shifts it to the point $(\frac{1}{3}, 0)$. The contraction and rotation are represented by multiplication by a matrix A, while the translation is represented by an additive vector $\mathbf{b}$ as follows:

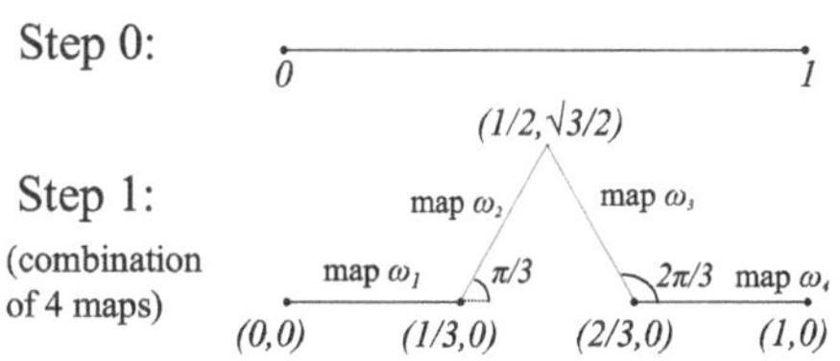

Figure 6.11: Combination of the four maps (ω_1–ω_4) used in the construction of the Koch curve. Illustration of the first step in the generation process.

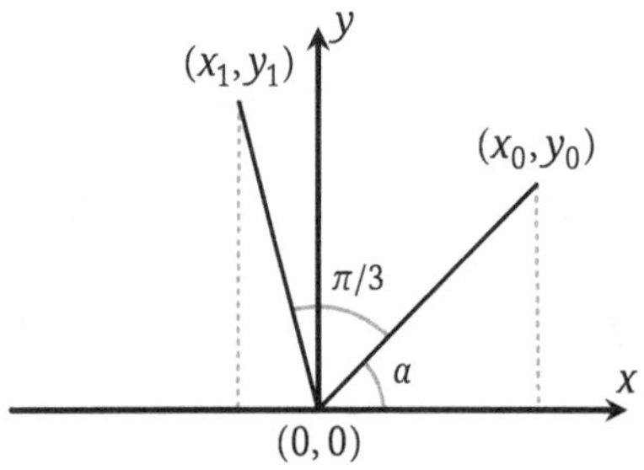

Figure 6.12: Finding the end point of the arbitrary unit vector after its rotation by $\pi/3$.

$$w_2 : \begin{pmatrix} x_{n+1} \\ y_{n+1} \end{pmatrix} = A \begin{pmatrix} x_n \\ y_n \end{pmatrix} + \mathbf{b}.$$

To find the matrix A in the definition of ω_2, we refer to Figure 6.12. If A_0 represents the rotation of the position vector of the point $(x_0, y_0) = (\cos\alpha, \sin\alpha)$ by an angle $\pi/3$ without changing its length, then the image (x_1, y_1) of the point (x_0, y_0) satisfies

$$x_1 = \cos\left(\alpha + \frac{\pi}{3}\right) = \cos\alpha\cos\frac{\pi}{3} - \sin\alpha\sin\frac{\pi}{3} = x_0\cos\frac{\pi}{3} - y_0\sin\frac{\pi}{3},$$

and

$$y_1 = \sin\left(\alpha + \frac{\pi}{3}\right) = \sin\alpha\cos\frac{\pi}{3} + \cos\alpha\sin\frac{\pi}{3} = y_0\cos\frac{\pi}{3} + x_0\sin\frac{\pi}{3}.$$

Combining the two formulas, we obtain

$$\begin{pmatrix} x_1 \\ y_1 \end{pmatrix} = \begin{pmatrix} \cos\frac{\pi}{3} & -\sin\frac{\pi}{3} \\ \sin\frac{\pi}{3} & \cos\frac{\pi}{3} \end{pmatrix} \begin{pmatrix} x_0 \\ y_0 \end{pmatrix}.$$

So,

$$A_0 = \begin{pmatrix} \cos\frac{\pi}{3} & -\sin\frac{\pi}{3} \\ \sin\frac{\pi}{3} & \cos\frac{\pi}{3} \end{pmatrix}.$$

The map w_2 shrinks the original object by a factor of three, that is,

$$A = \frac{1}{3}A_0,$$

and then it shifts the image horizontally so that the left-hand endpoint of the segment moves from $(0, 0)$ to $(\frac{1}{3}, 0)$. Therefore,

$$\mathbf{b} = \begin{pmatrix} \frac{1}{3} \\ 0 \end{pmatrix}.$$

Thus, w_2 is given by the map

$$\begin{pmatrix} x_{n+1} \\ y_{n+1} \end{pmatrix} = \frac{1}{3} A_0 \begin{pmatrix} x_n \\ y_n \end{pmatrix} + \mathbf{b} = \frac{1}{3} \begin{pmatrix} \cos\frac{\pi}{3} & -\sin\frac{\pi}{3} \\ \sin\frac{\pi}{3} & \cos\frac{\pi}{3} \end{pmatrix} \begin{pmatrix} x_n \\ y_n \end{pmatrix} + \frac{1}{3} \begin{pmatrix} 1 \\ 0 \end{pmatrix}.$$

To find w_3, we note that it involves rotation by $2\pi/3$, tree-fold shrinkage and shift along x-axis by 2/3. Following the procedure, used to find ω_2, we get

$$w_3 : \begin{pmatrix} x_{n+1} \\ y_{n+1} \end{pmatrix} = \frac{1}{3} \begin{pmatrix} \cos\frac{2\pi}{3} & -\sin\frac{2\pi}{3} \\ \sin\frac{2\pi}{3} & \cos\frac{2\pi}{3} \end{pmatrix} \begin{pmatrix} x_n \\ y_n \end{pmatrix} + \frac{2}{3} \begin{pmatrix} 1 \\ 0 \end{pmatrix}.$$

The map w_4, does not rotate the original object, but shrinks it three times and shifts it left by 2/3. So,

$$w_4 : \begin{pmatrix} x_{n+1} \\ y_{n+1} \end{pmatrix} = \frac{1}{3} \begin{pmatrix} 1 & 0 \\ 0 & 1 \end{pmatrix} \begin{pmatrix} x_n \\ y_n \end{pmatrix} + \frac{2}{3} \begin{pmatrix} 1 \\ 0 \end{pmatrix}.$$

In Step 1 we apply all four maps to the unit segment used as the starting point for constructing the Koch curve, that is, to all points (x_0, y_0) with $x_0 \in [0, 1]$ and $y_0 = 0$. In Step 2 we apply the same four maps to every point (x_1, y_1) obtained in the previous iteration, and this process is repeated indefinitely. The set of all possible points (x_n, y_n) as $n \to \infty$ forms the Koch curve.

6.7.4 Combining maps to build Sierpiński triangle

This is done in the same manner as in the case of the Koch curve. We find three maps such that each of them maps the unit triangle into twice smaller triangle. Furthermore, one map keeps the position of the obtained triangle, while the second and third shift it to reproduce Step 1 in the construction of the Sierpiński triangle. Following clues given in Figure 6.13, it is easy to check that these maps are given as

$$w_1 : \begin{pmatrix} x_{n+1} \\ y_{n+1} \end{pmatrix} = \frac{1}{2} \begin{pmatrix} 1 & 0 \\ 0 & 1 \end{pmatrix} \begin{pmatrix} x_n \\ y_n \end{pmatrix}$$

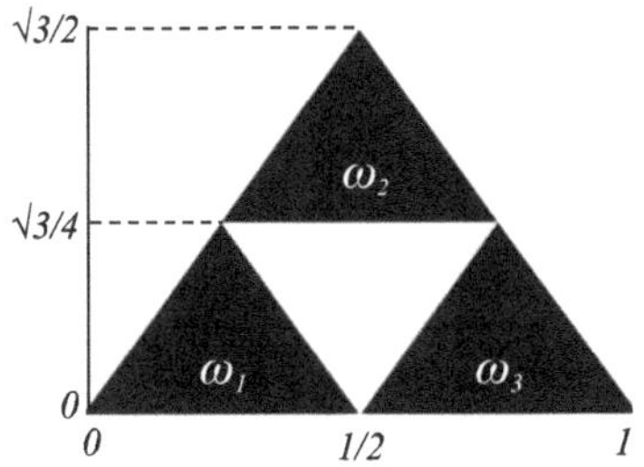

Figure 6.13: Combination of the three maps (ω_1–ω_4) used in the construction of the Sierpiński triangle. Illustration of the first step in the generation process.

$$w_2 : \begin{pmatrix} x_{n+1} \\ y_{n+1} \end{pmatrix} = \frac{1}{2}\begin{pmatrix} 1 & 0 \\ 0 & 1 \end{pmatrix}\begin{pmatrix} x_n \\ y_n \end{pmatrix} + \frac{1}{4}\begin{pmatrix} 1 \\ \sqrt{3} \end{pmatrix}$$

$$w_3 : \begin{pmatrix} x_{n+1} \\ y_{n+1} \end{pmatrix} = \frac{1}{2}\begin{pmatrix} 1 & 0 \\ 0 & 1 \end{pmatrix}\begin{pmatrix} x_n \\ y_n \end{pmatrix} + \frac{1}{2}\begin{pmatrix} 1 \\ 0 \end{pmatrix}.$$

6.8 Summary

In this chapter; we have introduced a range of fascinating geometric objects known as fractals and examined some of their key properties, including their geometric dimensions. The modern theory of fractals was developed largely during the second half of the twentieth century, and many aspects of it continue to evolve. Several notions of fractal dimension have been proposed, and in this chapter we have focused on only one of them: the capacity dimension. Although useful, it has significant limitations. More advanced concept—most notably the Hausdorff dimension [10]—provide a far richer and more robust framework, but their study requires measure theory, which would take us too far from the main direction of this text.

Chaotic regimes in dynamical systems are commonly associated with the formation of so-called strange attractors, which typically possess a fractal structure. We have used the period-doubling cascade to illustrate how a fractal attractor emerges as chaos develops in one-dimensional maps. In addition, we have shown that iterated maps can be used effectively to generate a wide variety of fractal objects.

Although fractals may appear to be sophisticated mathematical objects, they are frequently encountered in nature. In biology, fractal patterns arise in the branching structure of the bronchial tree in the lungs, in blood vessels, and in the growth patterns of ferns, broccoli, and many other organisms. In physics, fractal forms are seen in lightning paths, coastlines, galaxy distributions, snowflakes, and numerous other natural phenomena.

Fractals therefore provide a beautiful meeting point between geometry, dynamics, and nature. They reveal how simple rules can generate extraordinary complexity and how chaotic behavior can give rise to structures of remarkable order and elegance.

6.9 Exercises

Problem 6.1. Consider the logistic map given by iterations of the function

$$f(x) = Ax(1 - x),$$

where A and x are complex numbers ($A, x \in \mathbb{C}$). Find the regions of the 'complex A-plane' for which the map has

(i) A stable fixed point.

(ii) A stable period-2 solution.

Problem 6.2. Consider the map $x_{n+1} = f(x_n)$, where

$$f(x) = \begin{cases} 4x, & 0 \le x < \frac{1}{4}, \\ 4x - 2, & \frac{1}{2} \le x < \frac{3}{4}, \\ 0, & \text{otherwise.} \end{cases}$$

(i) Sketch $f(x)$. Find the fixed points of this system and show that they are unstable.
(ii) Consider the set S of initial points x_0 for which $x_n \nrightarrow 0$ as $n \to \infty$. Obtain a description of S and use it to find the capacity dimension of S.

Problem 6.3. A dynamical system on $[0, 1]$ is given by $x_{n+1} = f(x_n)$, where

$$f(x) = \begin{cases} 5x, & \text{for } 0 \le x \le \frac{1}{5}, \\ 5x - 2, & \text{for } \frac{2}{5} \le x \le \frac{3}{5}, \\ 5x - 4, & \text{for } \frac{4}{5} \le x \le 1, \\ 0, & \text{otherwise.} \end{cases}$$

(i) Sketch the function $f(x)$. Determine all fixed points of the system and show that each of them is unstable.
(ii) Let S denote the set of initial points x_0 for which $x_n \nrightarrow 0$ as $n \to \infty$. Provide a description of the set S, and use this description to compute the capacity dimension of S.

Problem 6.4. Consider a square subdivided into an 8×8 grid of smaller squares, as on a chessboard. Let Q be the operation that removes all black squares. Apply Q once to the unit square, leaving 32 subsquares. Then, apply Q again to each of the remaining subsquares, and continue iterating this procedure n times, with $n \to \infty$. Determine the capacity dimension of the resulting set.

6.10 Solutions to exercises

Solution to Problem 6.1.

(i) The fixed points of the map

$$x_{n+1} = Ax_n(1 - x_n)$$

are $X_1 = 0$ and $X_2 = 1 - 1/A$. Since $f'(x) = A(1 - 2x)$, we have

$$|f'(0)| = |A|$$

so $X_1 = 0$ is asymptotically stable if $|A| < 1$, i. e., inside the unit circle centered at the origin in the complex A-plane.

Also,

$$|f'(1-1/A)| = |2-A|$$

so $X_2 = 1 - 1/A$ is asymptotically stable for $|2 - A| < 1$, i. e., inside the unit circle centered at the point $A = 2$ in the complex A-plane.

(ii) The 2-cycle for the logistic map is such that

$$x_{n+2} = f(f(x_n)) = A[Ax_n(1-x_n)][1-Ax_n(1-x_n)].$$

The fixed points of this iterated mapping are X_1 and three more roots satisfying

$$A^3x^3 - 2A^3x^2 + (A^3 + A^2)x + (1 - A^2) = 0.$$

This equation can be factorized as

$$[Ax + (1-x)][A^2x^2 - (A^2 + A)x + (1 + A)] = 0$$

since $X_2 = 1 - 1/A$ is a fixed point of f and hence of $f \circ f$. The point $x = 0$ is also a fixed point of $f \circ f$, but the points X_3 and X_4 of a genuine 2-cycle (i. e., $f(X_3) = X_4$ and $f(X_4) = X_3$) satisfy

$$A^2x^2 - (A^2 + A)x + (1 + A) = 0. \tag{6.6}$$

Such fixed points are asymptotically stable when

$$\left|\frac{d}{dx}f(f(x))\Big|_{x=X_3}\right| = |f'(f(X_3))f'(X_3)| = |f'(X_4)f'(X_3)| < 1.$$

Hence, we require that

$$|A^2(1-2X_3)A(1-2X_4)| < 1,$$

or

$$|A^2[1 - 2(X_3 + X_4) + 4X_3X_4]| < 1. \tag{6.7}$$

From (6.6), we note that

$$X_3 + X_4 = \frac{A^2 + A}{A^2}$$

and

$$X_3X_4 = \frac{1+A}{A^2}.$$

From (6.7), we see that the roots must satisfy

$$|-A^2 + 2A + 4| < 1$$

or

$$|(A-1)^2 - 5| < 1.$$

Hence, we have

$$|A - 1 - \sqrt{5}||A - 1 + \sqrt{5}| < 1,$$

which corresponds to two disjoint ovals in the complex A-plane, oriented across the x-axis, with endpoints $(-1-\sqrt{6}, 0)$ and $(-1, 0)$ for the first one and $(3, 0)$ and $(1+\sqrt{6}, 0)$ for the second one. These ovals touch the two unit circles found earlier as stability regions for the fixed points.

If we continue in this way, adding the regions of the complex A-plane in which cycles of all possible periods are stable, we obtain the full geometric structure shown in Figure 6.14.

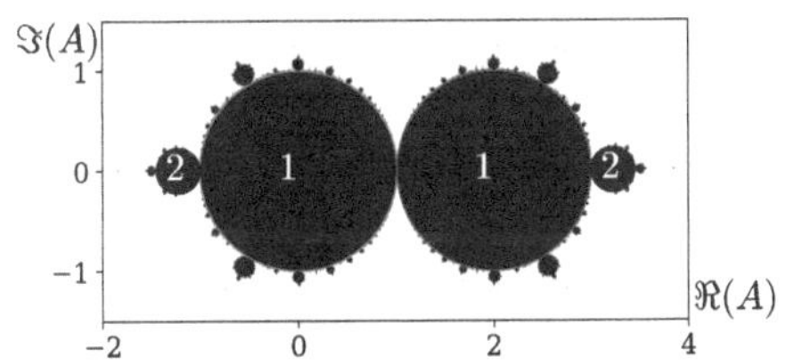

Figure 6.14: Regions of the complex A-plane for which the logistic map possesses an attracting fixed point (two unit circles, labelled '1') and an attracting 2-cycle (two disjoint ovals, labelled '2'). When additional regions corresponding to bounded solutions are included, the resulting set forms a fractal structure reminiscent of the Mandelbrot set shown in Figure 6.1.

Solution to Problem 6.2.

(i) The plot of the map-generating function is shown in Figure 6.15. As is evident from the graph, the map possesses two fixed points. One is the trivial fixed point $X_1 = 0$, and the other is obtained by solving

$$x = 4x - 2,$$

which yields the nontrivial fixed point $X_2 = \frac{2}{3}$.
Because the derivative of the map is constant, $f'(x) = 4$, and its value at each fixed point has a modulus greater than one. Consequently, both fixed points are unstable.

(ii) Let S_n denote the set of initial values x_0 for which the nth iterate satisfies $x_n \neq 0$. Concerning the sets obtained after each iteration, we note the following:

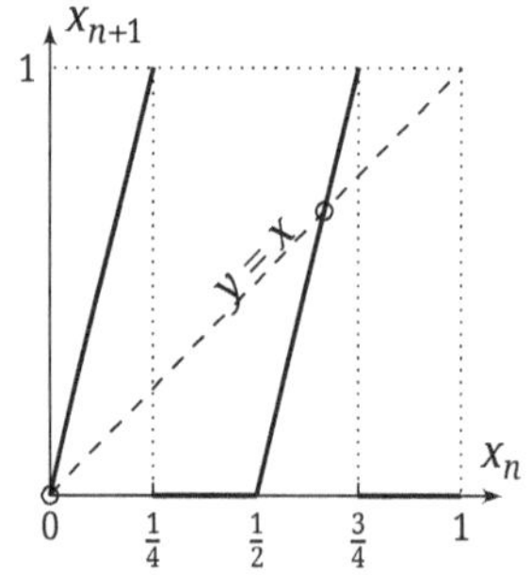

Figure 6.15: Graph of the function $f(x)$ given in Problem 6.2 (solid piecewise curve) together with the diagonal line $y = x$ (dashed). Their intersection points, marked by small circles, correspond to the fixed points of the map.

- As follows from Figure 6.15, the set S_1 consists of 2 intervals, each of length 4^{-1}.
- The set S_2 is obtained by mapping S_1 and consists of 2^2 intervals, each of length 4^{-2}.
- In general, S_n consists of 2^n intervals, each of length 4^{-n}.

The capacity (box-counting) dimension of S is, therefore,

$$d_c = \lim_{\varepsilon \to 0} \frac{\ln N(\varepsilon)}{\ln \varepsilon^{-1}} = \frac{\ln 2^n}{\ln 4^n} = \frac{n \ln 2}{n \ln 4} = \frac{\ln 2}{2 \ln 2} = \frac{1}{2}.$$

Thus, the capacity dimension of the set S is $d_c = 0.5$.

Solution to Problem 6.3.

(i) The sketch of $f(x)$ is shown in Figure 6.16. The function has slope 5 on each of its linear pieces, and it is equal to zero on the intervals [1/5, 2/5) and [3/5, 4/5). The fixed points of the system are

$$X_1 = 0, \quad X_2 = \frac{1}{2} \text{ (solving } x = 5x - 2), \quad X_3 = 1 \text{ (solving } x = 5x - 4),$$

as indicated in Figure 6.16.

In each case the derivative satisfies $|f'(x)| = 5 > 1$ so all three fixed points are unstable.

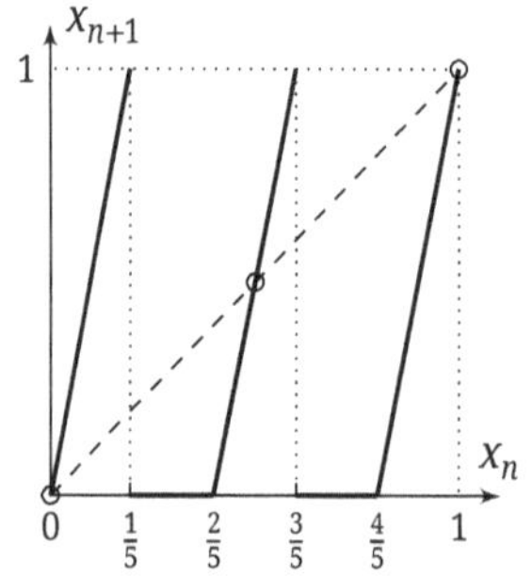

Figure 6.16: Graph of the function $f(x)$ given in Problem 6.3 (solid piecewise curve) together with the diagonal line $y = x$ (dashed). Their intersection points, marked by small circles, correspond to the fixed points of the map.

(ii) The set S of initial points for which $x_n \neq 0$ as $n \to \infty$ can be described as follows. From Figure 6.16, we see that the interval $[0,1]$ is divided into five equal subintervals. Let Q denote the operation that removes the second and fourth subintervals, leaving the remaining three. We now apply Q to each of these three surviving subintervals. Each of them is again subdivided into five equal parts: On the second and fourth parts, we have $f(f(x)) = 0$, while on the first, third, and fifth parts the function, $f(f(x))$ is linear with slope 25. Thus, after two iterations we are left with 3^2 subintervals. Repeating this procedure indefinitely removes all initial points x_0 for which $x_n = 0$ for some $n \to \infty$.
Hence, the remaining set S consists precisely of those initial points x_0 for which $x_n \neq 0$ for all sufficiently large n. This set is a self-similar Cantor-type discontinuum with scaling factor 5.
The quantities at each stage are summarized as

Level	1	2	⋯	n
Level scale ε	5^{-1}	5^{-2}	⋯	5^{-n}
Number of segments $N(\varepsilon)$	3	3^2	⋯	3^n.

The capacity (box-counting) dimension of S is, therefore,

$$d_c = \lim_{\varepsilon\to 0} \frac{\ln N(\varepsilon)}{\ln(1/\varepsilon)} = \frac{\ln 3^n}{\ln 5^n} = \frac{\ln 3}{\ln 5} \approx 0.6826.$$

Solution to Problem 6.4. Let us now imagine that you are a reasonably good chess player, and consider a square divided into 8×8 smaller squares, as on a classical chessboard. Unlike an ordinary chessboard, however, we focus on the set obtained by removing all the black squares. This leaves 32 white squares. We then repeat the same operation on each of these 32 remaining squares, applying the chessboard subdivision and removing the black squares at every stage.

We are clearly dealing with a self-similar structure (with scaling factor 8), so it is natural to compute its capacity dimension. The quantities at each iteration are summarized as

Level	1	2	⋯	n
Level scale ε	8^{-1}	8^{-2}	⋯	8^{-n}
Number of segments $N(\varepsilon)$	1	32	⋯	32^n.

The capacity (or box-counting) dimension of the resulting set is, therefore,

$$d_c = \lim_{n\to\infty} \frac{\ln N(\varepsilon)}{\ln(1/\varepsilon)} = \lim_{n\to\infty} \frac{\ln 32^n}{\ln 8^n} = \frac{5\ln 2}{3\ln 2} = \frac{5}{3}.$$

Index

https://doi.org/10.1515/9783111348339-007

Bibliography

[1] Edward A. Bender. *An Introduction to Mathematical Modeling*. Dover Publications, Mineola, NY, 2000.

[2] J. S. Berry and K. Houston. *Mathematical Modelling. Modular Mathematics Series*. Butterworth-Heinemann, London, 1995.

[3] William E. Boyce, Richard C. DiPrima, and Douglas B. Meade. *Elementary Differential Equations and Boundary Value Problems*, 12th edition. Wiley, 2022.

[4] A. S. Bratus', A. S. Novozhilov, and A. P. Platonov. *Dinamicheskie Sistemy i Modeli Biologii*. Fizmatlit, Moscow, 2010.

[5] Nicholas F. Britton. *Essential Mathematical Biology*. Springer Undergraduate Mathematics Series. Springer, London, 2003.

[6] Thomas H. Cormen, Charles E. Leiserson, Ronald L. Rivest, and Clifford Stein. *Introduction to Algorithms*, 4th edition. MIT Press, 2022.

[7] Robert L. Devaney. *An Introduction to Chaotic Dynamical Systems*, 2nd edition. Addison-Wesley, Redwood City, CA, 1989.

[8] Mitchal Dichter. *Student Solutions Manual for Nonlinear Dynamics and Chaos: With Applications to Physics, Biology, Chemistry, and Engineering*, 3rd edition. CRC Press, Boca Raton, 2024.

[9] Saber N. Elaydi. *An Introduction to Difference Equations*, 3rd edition. Springer, New York, 2005.

[10] Kenneth J. Falconer. *Fractal Geometry: Mathematical Foundations and Applications*, 3rd edition. John Wiley & Sons, Chichester, 2014.

[11] Morris W. Hirsch, Stephen Smale, and Robert L. Devaney. *Differential Equations, Dynamical Systems, and an Introduction to Chaos*, 3rd edition. Pure and Applied Mathematics. Academic Press, 2013.

[12] Yulij Ilyashenko and Sergei Yakovenko. *Lectures on Analytic Differential Equations*. Graduate Studies in Mathematics, volume 86. American Mathematical Society, Providence, RI, 2008.

[13] Anatole Katok and Boris Hasselblatt. *Introduction to the Modern Theory of Dynamical Systems*. Encyclopedia of Mathematics and its Applications, volume 54. Cambridge University Press, Cambridge, 1995.

[14] Yuri A. Kuznetsov. *Elements of Applied Bifurcation Theory*, 2nd edition. Springer, Cham, Switzerland, 2023.

[15] Albert C. J. Luo. *Bifurcation and Stability in Nonlinear Dynamical Systems*. Nonlinear Systems and Complexity, volume 28. Springer International Publishing, Cham, 2019.

[16] Benoît B. Mandelbrot. How long is the coast of Britain? Statistical self-similarity and fractional dimension. *Science*, 156(3775):636–638, 1967.

[17] J. D. Murray. *Mathematical Biology: I. An Introduction*, 3rd edition. Springer, 2003.

[18] Walter Rudin. *Principles of Mathematical Analysis*, 3rd edition. McGraw-Hill, New York, 1976.

[19] Steven H. Strogatz. *Nonlinear Dynamics and Chaos: With Applications to Physics, Biology, Chemistry, and Engineering*, 3rd edition. Chapman & Hall/CRC, 2024.

https://doi.org/10.1515/9783111348339-008

De Gruyter Series in Applied and Numerical Mathematics

Volume 11
Umurdin Dalabaev
Solution of Initial-Boundary Value Problems. Method of Moving Modes, 2025
ISBN 978-3-11-130439-7, e-ISBN (PDF) 978-3-11-130817-3, e-ISBN (EPUB) 978-3-11-130873-9

Volume 10
Cholmin Sin, Evgenii S. Baranovskii
Regularity Theory for Generalized Navier–Stokes Equations. Non-Newtonian Fluids with Variable Power-Law, 2025
ISBN 978-3-11-150876-4, e-ISBN (PDF) 978-3-11-150890-0, e-ISBN (EPUB) 978-3-11-150948-8

Volume 9
Habib Ammari, Bryn Davies
Metamaterial Analysis and Design. A Mathematical Treatment of Cochlea-inspired Sensors, 2023
ISBN 978-3-11-078404-6, e-ISBN (PDF) 978-3-11-078496-1, e-ISBN (EPUB) 978-3-11-078513-5

Volume 8
Zhi-Zhong Sun, Qifeng Zhang, Guang-hua Gao
Finite Difference Methods for Nonlinear Evolution Equations, 2023
ISBN 978-3-11-079585-1, e-ISBN (PDF) 978-3-11-079601-8, e-ISBN (EPUB) 978-3-11-079611-7

Volume 7
Roland Glowinski, Tsorng-Whay Pan
Numerical Simulation of Incompressible Viscous Flow. Methods and Applications, 2022
ISBN 978-3-11-078491-6, e-ISBN (PDF) 978-3-11-078501-2, e-ISBN (EPUB) 978-3-11-078505-0

Volume 6
Bruno Després
Neural Networks and Numerical Analysis, 2022
ISBN 978-3-11-078312-4, e-ISBN (PDF) 978-3-11-078318-6, e-ISBN (EPUB) 978-3-11-078326-1

Volume 5/1
Alexander V. Bobylev
Kinetic Equations. Volume 1: Boltzmann Equation, Maxwell Models, and Hydrodynamics beyond Navier–Stokes, 2020
ISBN 978-3-11-055012-2, e-ISBN (PDF) 978-3-11-055098-6, e-ISBN (EPUB) 978-3-11-055017-7

FSC
www.fsc.org
MIX
Papier | Fördert
gute Waldnutzung
FSC® C083411

Kolibri 360 GmbH
Ferdinand-Jühlke-Straße 7
99095 Erfurt, Deutschland
produktsicherheit@kolibri360.de